Properties and Applications of Glass

GLASS SCIENCE AND TECHNOLOGY

Glass Science and Technology 3

Properties and Applications of Glass

HAROLD RAWSON
Department of Ceramics, Glasses and Polymers,
University of Sheffield, England

ELSEVIER
Amsterdam — Oxford — New York   1980

ELSEVIER SCIENCE PUBLISHERS B.V.
Molenwerf 1
P.O. Box 211, 1000 AE Amsterdam, The Netherlands

Distributors for the United States and Canada:

ELSEVIER SCIENCE PUBLISHING COMPANY INC.
52, Vanderbilt Avenue
New York, NY 10017

First edition 1980
Second impression 1984

Library of Congress Cataloging in Publication Data
Rawson. Harold.
  Properties and applications of glass.

  (Glass science and technology ; 3)
  Bibliography: p.
  Includes indexes.
  1. Glass. I. Title. II. Series.
TA450.R33   620.1'44   80-21123
ISBN 0-444-41922-5

ISBN 0-444-41922-5 (Vol. 3)
ISBN 0-444-41577-7 (Series)

Printed in The Netherlands

PREFACE

Hilaire Belloc introduced his book "The Path to Rome" not with a preface but with a few pages under the heading "In Praise of This Book". Authors of scientific texts are expected to be far more modest, even apologetic. However no one needs to apologize for writing a book about glass (or "glasses" as some prefer, to emphasize the fact that the term embraces a very wide range of materials). Very few books have been written on glass compared with the number which deal with other important classes of materials, e.g. metals, ceramics and polymers.

Although the reader is not entitled to an apology, he or she may welcome an explanation of why this book came to be written and some indication of the author's intentions. These are relatively modest and, on the whole, practical. For the first part of my working life, some seventeen years, I was fortunate enough to work in a relatively large research laboratory in the British electrical engineering industry. The work involved a good deal of contact with both electrical and mechanical engineers, who were using a variety of glasses, especially in lamps and electronic valves. Each problem had its own special features, but throughout there was a continuous need to convey to engineering colleagues some of the general ideas and understanding which they needed if they were to use glass sensibly. Their interests have certainly influenced my selection of material and the emphasis which I have given to some topics. I hope that they and their successors will find the book useful.

For the last fifteen years I have probably been equally fortunate in working in a university department which has a tradition, lasting more than sixty years, of teaching Glass Technology to undergraduate and, more recently, postgraduate students. Any university teacher of science and technology must have a great deal of sympathy for his students. With the passage of time and as library shelves fill at an increasingly rapid rate, the teacher finds himself under increasing pressure to transmit more ideas and information about his subject. In the particular field covered by this book, one has at least to take notice of new applications (such as the use of glasses in lasers and fibre optic communication systems), of new types of glasses (the metallic glasses) and of new ideas, many of which have been injected into the subject by physicists, who in increasing numbers are turning their attention from the solid state physics of crystals in order to develop the physics of amorphous solids.

Those of us who are teaching in this field have no general and wide ranging textbook, written at a relatively elementary level, to which we can refer our students. I hope they (that is the students) will find this book useful. It deals with recent developments only in outline. The intention has been rather to concentrate on those aspects of the properties of glasses which are most important in current practical applications. It should give students some insight into what I consider to be "mainstream" glass technology.

A third group of people who may find this book useful are my friends in the glass manufacturing industry. They are in-

credibly busy people who have to cope with a wide range of problems, the more pressing rarely being technical in nature. They
have even less time than their academic colleagues for what
ought to be called "running after the literature" rather than
the usual "keeping up with". They may occasionally find time
to thumb through the following pages and pick up an idea here
and there.

Although my aims have been practical in nature and designed
to help technologists rather than scientists, I hope that occasionally I have given the impression that the study of the glassy
state of matter is an interesting and certainly a very challeng-
ing activity. I also hope that I have given sufficient references, especially to review articles and other books, to make it
easy for the reader to develop a knowledge of glasses in a way
which suits his or her own particular interests.

The authors of many scientific texts are inclined to apologize to their families for what they consider to be the depriva-
tion resulting from their being busy about their self-imposed
tasks. One suspects that quite often the family is only too
happy that at least one member is out of the way and not creat-
ing a nuisance. However I offer half-hearted apologies to my
sons for not helping more with their mathematics homework and
none at all to my daughter who is a geographer. I am grateful
to my wife for many things, but at present for tolerating an
excessive amount of paper about the house, for her help with
some of the typing and for her argumentative approach to some
of my English. I do not always agree with her. Consequently
errors and solecisms in the use of this difficult language are
entirely my responsibility.

Finally I wish to record my very sincere thanks to Mrs M.
Hodgins of the Department staff who has cheerfully undertaken
the task of producing a camera-ready copy at least approximately
to the demanding standard set by the publisher.

Sheffield   H. Rawson
May 1980

CONTENTS

ACKNOWLEDGEMENTS

I am most grateful to the following for permission to use figures and tables: Society of Glass Technology for Figures 11,
23, 25, 27, 28, 29, 33, 34, 57, 113, 114, 116, 119, 139, 140,
159, 161, 164, 165, 166; American Ceramic Society for Figures
3, 8, 9, 13, 20, 24, 30, 31, 32, 37, 69, 77, 92, 93, 115, 127,
129, 141, 156, 160, 162, 163; American Chemical Society for
Figure 2; American Institute of Physics for Figures 22, 66, 67,
68, 71, 120, 121; British Ceramic Society for Figure 122;
Deutsche Glastechnische Gesellschaft for Figures 87 and 88;
American Society for Testing of Materials for Figures 63, 82,
83; Ceramic Society of Japan for Figure 117; Butterworth and
Company Limited for Figures 109, 111, 112 from Bates (1962) in
"Modern Aspects of the Vitreous State", Vol. 2, Ed. J.D.
Mackenzie; North Holland Publishing Company for Figures 91, 98,
105, 157, 158, 168, 169; Taylor and Francis Limited for Figures
130, 145, 146, 147, 148, 149, 150 from Contemporary Physics and
for Figure 135 from "The Physical Properties of Glass" by D.G.
Holloway, Wykeham Publications Limited; British Glass Industry
Research Association for Figure 73; Columbia University Press
for Figure 128; Marcel Dekker Inc., N.Y., for Figures 118 and
123 from "Glass Structure and Spectroscopy" by J. Wong and C.A.
Angell (1976); John Wiley and Sons Limited for Figure 65 from
"Amorphous Materials" Ed. R.W. Douglas and B. Ellis (1972);
Academic Press Limited for Figures 132, 133, 137, 138 from
Hughes and Isard (1972) in "Physics of Electrolytes", Vol. I,
Ed. J.H. Hladik and for Figures 5 and 7 from "Glass Ceramics"
by P.W. McMillan (1964), Copyright by Academic Press Inc.
(London) Limited; Springer-Verlag for Figures 38, 39, 89, 126
from "Glas. Natur, Struktur und Eigenschaften" 2nd Edn. by H.
Scholze; Sijthoff and Noordhoff for Figure 70; Pergamon Press
Limited for Figures 124, 134, 136, 142, 144, 151, 152, 153, 154
from "Progress in Ceramic Science", Vol. 3, Ed. J.E. Burke
(1963); Plenum Publishing Corporation for Figures 58, 59, 60,
61 from Ernsberger (1962) in "Advances in Glass Technology"
Vol. I, and Figures 72 and 143 from "Introduction to Glass Science", Ed. L.D. Pye, H.J. Stevens and W.C. La Course (1972);
Associated Book Publishers Limited for Figure 110 from "An In-

troduction to Transition Metal Chemistry" by L.E. Orgel, Methuen (1966); McGraw Hill Book Company for Table VIII and Figure 46 from "Glass Engineering Handbook", 2nd Edn. by E.B. Shand (1958) and Figure 95 from "Modern Theory of Solids" by F. Seitz (1940); Corning Glass Works, Corning, N.Y., for Figures 6 and 131; Schott and Genossen, Jenaer Glasswerk for Figures 99 and 104.

CHAPTER 1

SOME ASPECTS OF THE NATURE OF INORGANIC
GLASSES

A. Definition of the Term "Glass"

One of the most frequently quoted definitions of the term
"glass" is that proposed in 1945 by the American Society for
Testing Materials: "Glass is an inorganic product of fusion
which has cooled to a rigid condition without crystallizing.".
This definition is satisfactory for those glasses with which we
are most familiar. It includes the materials used to make windows, glass containers, camera lenses, glass ovenware, lamp envelopes, etc. However, the A.S.T.M. definition is unnecessarily
restrictive since many organic materials form glasses, e.g. glycerol. Also, because we now know considerably more about materials than we did in 1945, we have to recognize that some materials made by methods other than cooling a melt have a claim to
be included. Thus, non-crystalline solids can be made by de-
position from the vapour phase, or by sputtering in a low pressure system, and these have the same chemical composition as,
and apparently identical properties to, glasses produced by
cooling from the melt. To meet this difficulty, a committee of
the U.S. National Research Council recently proposed a more general definition (Wong and Angell, 1976; p.36). Although this
cannot be fully understood without the background to be presented shortly, it will be quoted here: "Glass is an X-ray amorphous material which exhibits the glass transition, this being
defined as that phenomenon in which a solid amorphous phase exhibits with changing temperature a more or less sudden change
in the derivative thermodynamic properties, such as heat capacity and expansion coefficient, from crystal-like to liquid-like
values.". Some authors are apparently prepared to go further
and apply the adjective "glassy" or "vitreous" to almost any
amorphous solid. This indiscriminate practice is to be discour-
aged. The characteristic features of the glassy state are well
understood and are usually easily recognizable. The term "glass"
should be applied only to materials which show these features.

Rather than spending more time in agonizing over the niceties
of definition or in considering borderline cases, it is better
to give at once an account of the behaviour of the most characteristic glass-forming materials.

Most elements and compounds when molten have a viscosity
about the same as that of water (10^{-2} Pa s). On cooling the
melt, crystallization occurs very rapidly at, or a little below
the freezing point. There are, however, a few materials which
form melts which are considerably more viscous. The high viscosity indicates that the atoms or molecules in the melt are
not so easily moved relative to one another by applied stresses.
On cooling below the freezing point, crystallization does occur,
but at a significantly lower rate than in the materials of the
first group. The process of crystallization involves structural
changes, i.e. the re-arrangement of atoms relative to one another. In simple terms, the relatively high viscosity of the
melt and the low rate of crystallization are both consequences
of the greater resistance to atomic re-arrangement encountered
in these materials.

If the crystallization rate is low enough, it is possible
to go on cooling the melt below the freezing point without crystallization taking place. As the melt cools, its viscosity continues to increase. This viscous liquid below the freezing
point is a supercooled liquid. It is incorrect to refer to it
as a glass. Further cooling results in the viscosity rising to
such a high value that the mechanical properties of the material
are closely similar to those of an ideal elastic solid. The
viscosity will then be at least 10^{12} to 10^{13} Pa s. This solid
material is a glass.

B.   The Transformation Range

The volume-temperature diagram shown in Fig.1 is useful in discussing the transformation from a supercooled liquid to a glass.
If the melt crystallizes on cooling, this is usually accompanied by a marked increase in density at the melting point, T_f.
No such change occurs if the melt supercools. The volume decreases along the line, be. The decrease in volume on cooling
is due partly to the decreasing amplitude of atomic vibrations,

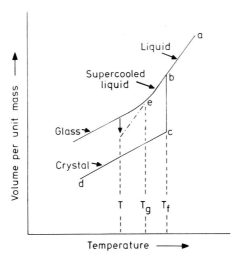

Fig.1. Relation between the glassy, liquid and solid states.

and partly to changes in the structure of the melt which result in it becoming more compact as the temperature falls. At temperatures near T_f these structural changes can occur very rapidly and will *appear* to occur instantaneously following any change in the temperature of the material. As the viscosity increases with falling temperature, the structural changes occur increasingly slowly until eventually the viscosity becomes so high that no further such changes are possible. A decrease in slope is then found in the volume-temperature curve (point e). With a further fall of temperature, the decreasing volume is due almost entirely to the decreasing amplitude of the atomic vibrations.

The temperature at which the change in slope occurs is called the transformation temperature or glass transition temperature, T_g. Only below T_g is it correct to describe the material as a glass. The change from supercooled liquid to glass which may be considered as taking place at this temperature is not a sudden one, nor is T_g a well-defined temperature for any particular glass. Indeed the term "transformation range" is used more frequently than "transformation temperature". The temperature at which the change in slope occurs is found to decrease as the rate of cooling is decreased. Also, if the glass is held at the temperature T, a little below T_g, its volume decreases slowly until it reaches a point on the dotted line, which is an extrapolation of the contraction curve of the super-

cooled liquid. The rate of change of volume decreases as the
dotted line is approached, i.e. as the structure of the glass
approaches an equilibrium "configuration" which is characteristic of the supercooled melt at the temperature T. This equilibrium configuration has a lower free energy than other liquid-
like structures or configurations, but it is not, of course,
that arrangement of the atoms or molecules in the material which
has the lowest possible free energy at the temperature T. At
any temperature below the melting point T_f, the arrangement of
the lowest possible free energy is that of the crystalline material. However at temperatures significantly below T_g, the rate
at which the liquid-like glass structure can change to the regular arrangement characteristic of the crystalline material is
infinitely slow. There is no reason to believe that the first
glass articles which were made four or five thousand years ago
are more crystalline now than on the day they were made. A
glass in its equilibrium configuration at a particular temperature T is said to be in a state of *metastable equilibrium*. The
reader unfamiliar with this concept may find it helpful to consider the potential energy of a house brick supported on a horizontal surface. The state of lowest energy is when the brick
is lying on its side, when its centre of gravity is as close to
the supporting surface as possible. The equilibrium configuration of a glass is analogous to the brick standing on one end.
The brick will stand for ever unless it is pushed over far
enough for its centre of gravity to pass through the vertical
plane which includes the line of contact between one edge of
the brick and the supporting surface. To tilt the brick over
into this unstable position, it is necessary to expend work,
because the initial tilting involves raising its centre of
gravity.

In discussing the changes in configuration and physical
properties which occur when a glass is heat treated in the vicinity of the transformation range, the concept of a"fictive
temperature" has proved to be of some value. This concept was
introduced by A.Q. Tool in the early 1920's to provide a numerical measure of the configuration of the glass and of the degree
to which that configuration departs from the equilibrium configuration when the glass is at a specified, measured temperature,

T. If we imagine a glass being brought to equilibrium at a temperature T´, which is above T_g, and then cooled infinitely quickly to a real temperature T, significantly below T_g, there will be no time for the configuration to change during the temperature change. The glass at T will have a configuration identical to the equilibrium configuration at T´. The temperature T´ is the fictive temperature of the glass at the temperature T. If, on the other hand, the glass is cooled slowly from T´, its fictive temperature will lie somewhere below T´. In principle, the fictive temperature can be determined experimentally by measuring a property of the glass at T and comparing this with values of the same property of samples cooled extremely rapidly from a series of temperatures below T´. Unfortunately, it is found that the fictive temperature determined in this way depends to some extent on which physical property is chosen to determine it. It appears that more than one number or parameter is required to specify the configuration of a glass specimen. This subject still attracts a considerable amount of study and discussion.

Although the volume-temperature diagram is most frequently used in introducing discussions of the transformation range, the rate of change with temperature of other properties would serve equally well. Figure 2 shows the temperature variation of the heat content per gram of glassy and supercooled glycerol in the vicinity of the transformation range.

Below 185°K the results depend upon the time allowed to bring the specimen to equilibrium. Above this temperature the heat energy supplied to the material contributes partly towards increasing the kinetic energy of the molecules and the atoms within each molecule, and partly towards supplying the energy required to produce a more open configuration. At temperatures well below T_g, no configurational changes occur when the temperature is changed and the heat content accordingly changes less rapidly with temperature.

From the foregoing account it should be clear that the properties of a glass measured at room temperature will depend upon the rate at which it was cooled through the transformation range. When exact control of glass properties is required, these heat treatment effects are of considerably practical im-

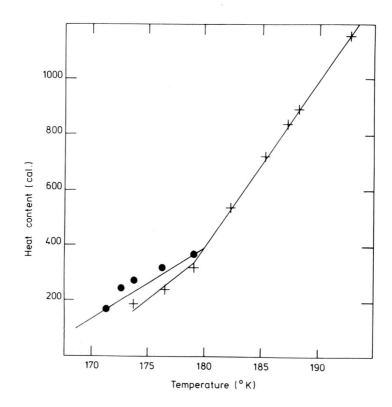

Fig.2. Heat content of supercooled glycerol. The crosses are for short time and the circles for long time experiments (Oblad and Newton, 1937).

portance. A number of examples of the effects of cooling rate on glass properties will be described in later chapters.

C. The Chemical Nature of Inorganic Glasses

Glass formation has been observed in a very large number of inorganic systems. A review of these has been made by the present author (Rawson, 1967). It is sufficient here to give a brief indication of the systems in which glass formation occurs emphasizing in particular those which are of practical importance.

Table I includes the few simple inorganic substances known to form glasses. Of these, the oxides SiO_2, B_2O_3 and P_2O_5 are by far the most important. Nearly all commercial glasses are based on these oxides.

TABLE I

Simple glass-forming substances

Substance	Melting Point (°K)
Se	490
SiO_2	1996
B_2O_3	723
GeO_2	1389
As_2O_3	551-582
P_2O_5	695-853
$ZnCl_2$	591
BeF_2	823
As_2S_3	583
Toluene	178
Ethanol	156
Glycerol	186

Vitreous silica is an important material in its own right and is widely used to make chemical ware, lamp envelopes, and in the manufacture of semiconductors.

It we consider melts made from two or more components, the range of glass-forming systems is enormously extended. By melting any of the oxides in Table I with a second (usually basic) oxide, many glass-forming systems may be formed. Thus glasses can be made over a wide range of compositions in all the alkali-silicate systems: Li_2O-SiO_2, Na_2O-SiO_2, K_2O-SiO_2, etc. The range of glass-forming compositions is continuous from silica itself up to about 50 mol.% of the alkali oxide. Extensive regions of glass formation are also found in many borate and phosphate systems. One can also make glasses over wide regions of composition in several halide systems.

Glasses based on the elements S, Se and Te, in combination with other elements, especially As, Sb, Ge and Si, have been intensively studied in recent years on account of their interest- ing electrical and optical properties. This group is referred to as the chalcogenide glasses.

There are a number of interesting glass-forming systems

based on the oxides Al_2O_3, TeO_2, V_2O_5 and a few others. None
of these oxides alone will form a glass. However, many aluminate, tellurite and vanadate glasses can be made by melting Al_2O_3,
TeO_2 and V_2O_5 respectively in suitable proportions with a second
oxide or mixture of oxides. The aluminate glasses transmit to
somewhat longer wavelengths in the infra-red than silicate
glasses and this has resulted in their use to make nose cones
for heat-seeking missiles. The tellurite glasses are interesting on account of their high values of refractive index and dielectric constant, whilst the vanadate glasses are semi-conductors.

Other glasses can be made from mixtures of ionic salts, e.g.
in the system KNO_3-$Ca(NO_3)_2$, and even from aqueous solutions.

Finally, one of the most exciting discoveries of recent
years in glass technology is that many metallic alloys can be
produced in glassy form. To prevent the melts from crystallizing it is necessary to cool them much more rapidly than the
silicate glass-forming melts. Consequently, in order to obtain
a high rate of heat extraction throughout the thickness of the
material, metallic glasses can be made only as thin ribbons or
foils. However, this does not imply that metallic glasses are
not likely to be of technological importance. The properties
vary over a very wide range. Some have an extremely high tensile strength combined with remarkably high ductility. Other
compositions have a high corrosion resistance, a property which
seems to be explainable in terms of the absence of grain boundaries. Yet others have interesting magnetic properties. They
are magnetically very "soft", i.e. they may be magnetized very
easily by the application of small magnetic fields.

Useful reviews of the properties and compositions of metallic glasses have been published by Gilman (1975), Matsumoto and
Madden (1975), Takayama (1976) and Davies (1976). All these
materials, like their crystalline counterparts, are excellent
conductors of heat and electricity. Thus one must abandon the
commonly held view that "glasses are good insulators". General
statements about the properties of "glass" are now almost mean-
ingless.

Most commercially important glasses are oxide glasses and
silica is almost always a major component. A number of commer-

cial compositions are given in Table II. It will be noted that

TABLE II

Some commercial glass compositions (minor constituents are not given)

Application	Weight per cent									
	SiO_2	Al_2O_3	B_2O_3	MgO	CaO	BaO	ZnO	PbO	Na_2O	K_2O
Containers	72.2	1.9		1.5	9.6				14.6	
Window glass	72.0	1.3		3.5	8.2				14.3	
Lamp bulbs	71.5	2.0		2.8	6.6				15.5	1.0
Lead crystal	56.0							29.0	2.0	13.0
Tungsten sealing	75.5	2.6	16.0						3.7	1.7
Sodium vapour resistant	5.5	17.5	16.0		9.51	52.0				
Solder glass	5.0		17.0				14.0	64.0		

The reader is referred to Scholes (1975) and Tooley (1971) for a more complete account of commercial glass compositions.

most of them contain many components. These composition have usually been arrived at after many years of development, partly by systematic experiment, but more commonly by *ad hoc* adjustment. Each composition represents a compromise between several, often conflicting, requirements. These include: ease and economy of melting and fabrication; adequate chemical durability; and the need for the glass to have specified values of particular physical properties, e.g. refractive index, or thermal expansion.

D. Devitrification

1. The Liquidus Temperature

It will be obvious from a reading of the preceding paragraphs that single-component glass-forming melts, like other liquids, will crystallize if held for a sufficient time at temperatures which are below the melting point, but which are still high enough for structural re-arrangements to occur at a measurable rate. Some melts crystallize extremely slowly, boric oxide and a melt of the composition of soda felspar, $Na_2O.Al_2O_3.6SiO_2$, being outstanding in this respect.

Most glass-forming melts show some sign of crystallization (devitrification) if heated just below the liquidus temperature for a period of time ranging from minutes to hours. The liquidus temperature is the temperature below which a single liquid phase is no longer thermodynamically stable. If a melt consisting of two or more components is cooled to a temperature just below the liquidus, it does not, in general, crystallize as a whole. When equilibrium is reached, it usually consists of a mixture of a solid crystalline phase and a liquid. The liquidus temperature varies with temperature in a complicated way. Figure 3 shows this for part of the system Na_2O-SiO_2. This is

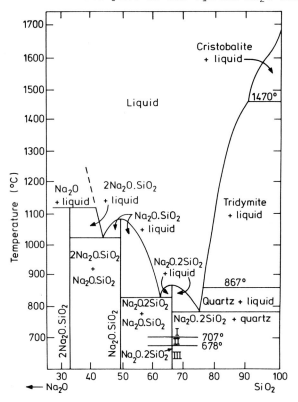

Fig.3. Phase diagram of part of the Na_2O-SiO_2 system.

the phase diagram, which shows not only the variation of liquidus temperature with composition, but also the crystalline phases which form. Within the region of glass formation there are two compounds: sodium metasilicate, $Na_2O.SiO_2$ (M. Pt. 1088°) and sodium disilicate, $Na_2O.2SiO_2$ (M. Pt. 874°).

The region of most interest is that between the composition of the disilicate and silica. In this region there is a composition at which the liquidus temperature is a minimum (a eutectic composition). This is the composition 74 w.% SiO_2, 26 w.% Na_2O. The very rapid decrease of liquidus temperature on adding Na_2O to silica is particularly noteworthy. Whilst temperatures of about 2000°C are required when making vitreous silica articles starting from pure sand or quartz, it would be possible to make articles from a soda-silica melt containing 25% Na_2O using melting temperatures as low as 1200°C. However, such articles would be of little practical value because the soda-silica glasses are rapidly attacked by atmospheric moisture. It is necessary to introduce other oxides, such as CaO and Al_2O_3, to make a glass of sufficient chemical durability. These additions increase the liquidus temperature and also the viscosity of the melt, so that a commercial soda-lime-silica glass, of the type used to make containers or flat glass, has a liquidus temperature of about 950°C and temperatures in the range 1400-1500°C are needed to produce a melt which will become sufficiently homogeneous and free from bubbles in an acceptable time.

The phase diagrams of many systems of interest to the glass technologist have been determined and are readily available in the invaluable compilation published by the American Ceramic Society (Levin et al. 1964, 1969, 1975).

2. Nucleation and Crystal Growth

The process by which optically visible crystals form in a melt below the liquidus temperature requires the existence of nuclei which act as centres from which larger crystals can grow. These nuclei are crystalline regions, often of sub-microscopic size. They may form spontaneously in the melt if its temperature is sufficiently below the liquidus temperature, in which case we speak of homogeneous nucleation. For reasons which will be described in a little more detail later, small percentages of certain materials are sometimes added to the glass batch; these being materials which readily precipitate from the melt

when it is cooled and form a very high density of nuclei. The
crystals which grow from the melt then have a composition which
differs radically from that of the nuclei. Finally, there are
situations in which crystals grow from foreign material inadver-
tently present in the melt; such as fragments of refractory,
particles of incompletely melted material, or particles of dust
on the melt surface. In these situations, where the nuclei are
not formed spontaneously from the major constituents of the melt
but can be regarded as foreign material, present either by acci-
dent or design, we speak of heterogeneous nucleation. A form of
heterogeneous nucleation which is frequently observed when a
piece of glass is heated in air to a temperature above the transformation range, is one in which crystals grow from nuclei on
the glass surface. The crystallization then proceeds by the
crystals, often needle-like in shape, growing inwards from the
surface. Usually the crystals are so numerous that they eventually form a white opaque layer covering the surface. This
detracts from the appearance of the glass and also reduces its
strength. Such devitrification is regarded as a defect. However
deliberately induced and controlled nucleation and crystal growth
in glass melts is now widely used to produce a variety of useful
materials, ranging from the gold ruby glasses which have been
known for centuries, to the class of materials called glass-
ceramics which have been developed only during the past twenty
years or so.

3. Mechanisms of Nucleation

There is space here to deal only briefly with nucleation
mechanisms. The reader is referred elsewhere for more complete
accounts (Fine, 1964). What follows is confined entirely to
homogeneous nucleation in a single-component system.

There is a decrease in free energy of ΔG per mole when the
liquid crystallizes. If one considers the effect only of this
bulk property of the material, the decrease in free energy in
forming a spherical crystalline region of radius r is
$4\pi r^3 . \Delta G/3V_m$, where V_m is the molar volume. However, one must
also take into account the contribution of the surface energy
of the interface between the crystal and the surrounding liquid.
If this is σ per unit area, the net change in free energy on

forming the crystal is

$$W = 4\pi r^3 \cdot \Delta G/3V_m + 4\pi r^2 \sigma. \qquad (1)$$

Figure 4 shows the variation of W with r.

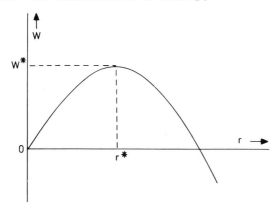

Fig.4. The free energy change in nucleus formation.

When r is less that r*, W is increasing with r. Consequently the free energy of the system will decrease if the crystalline region remelts. These very small, thermodynamically unstable crystalline regions are termed "embryos". They form
momentarily as a result of random fluctuations in the structure
of the liquid, but they disappear as quickly as they are formed.
An embryo will be a stable nucleus if a fluctuation occurs which
produces a crystalline region of radius greater than r*. The
value of r* at which W passes through a maximum is determined in
the usual way by differentiating equation (1) with respect to
r and setting the result equal to zero. This gives:

$$r* = -2\sigma V_m/\Delta G \qquad (2)$$

The increase in free energy, W*, associated with the formation
of a nucleus of this critical size is obtained by substituting
this value of r* in equation (1). Hence:

$$W* = 16\pi\sigma^3 V_m^2/3 \cdot \Delta G^2. \qquad (3)$$

It should be obvious, at least intuitively, that the formation
of a stable nucleus, since it involves an increase in free energy, is a relatively rare event both in time and in the space
occupied by the liquid. However, once a critically sized nucleus has been formed, its further growth will result in a decrease in free energy. Considerations based on statistical

thermodynamics show that the probability of thermal fluctuations leading to an increase in free energy equal to or greater than W* is given by the Boltzmann equation:

$$p(W^*) = \exp(- W^*/RT) ,\qquad\qquad (4)$$

where R is the gas constant. Thus the rate at which nuclei form will be proportional to $p(W^*)$. At the melting point, T_m, ΔG is zero. Equation (3) shows that W* is then infinitely large and thus $p(W^*)$ and the rate of nucleation will be zero. It may be shown that, under certain simplifying assumptions, $\Delta G = - \Delta H_f (T_m-T)/T_m$ where ΔH_f is the heat of fusion. If this expression is substituted in equation (3) and the resulting expression for W* in equation (4), an equationfor $p(W^*)$ is obtained which shows that $p(W^*)$ increases rapidly with decreasing temperature below T_m. However, the formation of a crystalline nucleus involves some structural re-arrangement in the melt, requiring that the atoms and molecules have sufficient thermal energy to overcome the energy barriers which impede these rearrangements. The rate of structural rearrangement is proportional to an Arrhenius rate factor, $\exp(- \Delta G_D/RT)$, where ΔG_D is the activation energy of the rearrangement process. Since activation energies usually vary only relatively slowly with temperature, the rate factor decreases as T decreases.

To sum up: the rate at which nuclei form depends upon two factors, a thermodynamic factor and a kinetic factor. The thermodynamic factor increases and the kinetic factor decreases as the temperature is reduced. The final equation for the rate of nucleation, I, is:

$$I = A.\exp(- W^*/RT).\exp(- \Delta G_D/RT) ,\qquad (5)$$

where A is a constant. A plot of I against T has the shape shown in Fig.5. Experimental measurements of the variation of I with T give curves of this shape, but detailed comparisons of theoretical and experimental values of the nucleation frequency are not usually possible, because it is difficult to obtain the values of some of the physical parameters which determine I, e.g. values of the interfacial energy, σ, and the activation energy, ΔG_D. (However, see Heady and Cahn, 1973.) If one assumes that the lowest nucleation rate which can be observed experimentally is about 1 cm^{-3} s^{-1}, then no nucleation will be observed until the melt has been cooled to some temperature

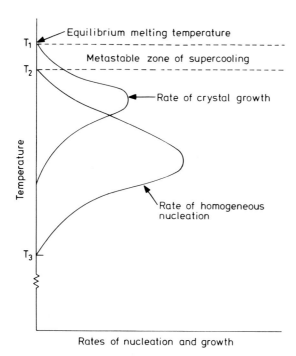

Fig.5. Effect of temperature on rates of nucleation and crystal growth.

measurably below T_f. This kind of behaviour is observed experimentally. No nucleation is detected until the melt has been supercooled to the temperature T_c, below which the nucleation rate increases rapidly with falling temperature. T_c is called the critical nucleation temperature. By inserting reasonable values for the physical parameters in equation (5), it is found that T_c is sensitive to both changes in σ and ΔG_D, decreasing as these two quantities increase. The values of r* are usually submicroscopic and values of I at high values of supercooling have usually been measured by holding the melt for some time at the nucleation temperature and then giving the material a second heat treatment at a higher temperature to allow the nuclei to grow to a detectable size.

4.   The Kinetics of Crystal Growth

Once a nucleus has been formed by the mechanism of homogeneous nucleation, or if the melt contains nuclei of foreign

material, crystals of easily measurable size can grow. A simple
theoretical analysis of the mechanism of crystal growth, due to
Turnbull and Cohen (1958), leads to the following equation for
the growth rate:

$$u = a_0 \nu . \exp(- \Delta G'/RT) . (1-\exp(\Delta G/RT)) \qquad (6)$$

ΔG is the decrease in free energy per mole when the liquid crystallizes and is expected to have the same value for macroscopic
crystals as for submicroscopic embryos (an expectation which may
not be fulfilled in practice). $\Delta G'$ is the activation energy, or
kinetic energy barrier, which must be overcome for an atom, molecule, or some other structural unit, to detach itself from the
liquid structure and attach to the surface of the growing crystal.
a_0 is a distance of the order of the interatomic distance and ν
is the frequency of thermal vibrations in the material. The
general form of the variation of u with T is also shown in Fig.5
and experimental results always show this type of variation.
Again, it is difficult to make comparisons between theoretical
and experimental values of u, largely because of the difficulty
in identifying the structural changes involved in the process of
crystal growth. This is particularly so if one is interested in
the practically important glasses which contain many components.
As Turnbull and Cohen point out, there is no reason to expect
$\Delta G'$ in equation (6) to be identical to ΔG_D in equation (5). The
structural mechanisms involved in forming a critical nucleus may
be quite different from those which determine macroscopic crystal
growth.

If one assumes that the lowest measurable growth rate is
about 1 μm hr^{-1}, and inserts reasonable values for ΔG and $\Delta G'$
in equation (6), it is found, as for nucleation, that the melt
must be cooled significantly below T_m before the rate of crystal
growth becomes measurable. However, the maximum growth rate
occurs at a much smaller degree of supercooling than that at
which the rate of nucleation is a maximum. This is fortunate
so far as glass formation is concerned. It will be easier to
form a glass if the maximum value of the growth rate is low relative to the rate at which heat can be extracted from the material. Unfortunately it is not possible at present to predict,
from the chemical composition of the melt alone, the values of
the physical parameters which determine u and I although, for

simple substances, some of these parameters may be estimated
from other properties of the melt.

The experimental measurement of crystal growth rates is
relatively easy in glass-forming silicate melts and it is frequently a matter of considerable practical importance to have
data on the variation of u with temperature. The crystals grow
slowly even at the temperatures at which the crystal growth is
most rapid. Very often the crystals are needle-like, so that
one can simply measure the length of crystals after various
times of heat treatment at a particular temperature. The literature contains much information on the effects of chemical composition on the rate of crystal growth in systems of technological importance. Good accounts of experimental techniques have
been given by Grauer and Hamilton (1950) and by Milne (1952).

5. The Effect of Glass Composition on the Rate of Devitrifi-
cation

The value of the maximum rate of crystal growth may change
markedly with even quite small changes in glass composition.
Also the addition of further constituents may greatly affect the
ease with which a glass can be made.

Table III shows how the maximum crystal growth rate varies
with composition in the system Na_2O-SiO_2.

Note the very large increase in growth rate on adding a
small percentage of Na_2O to SiO_2. Small additions of other
alkali or alkaline earth oxides have a similar effect. This is
the reason why a piece of silica glass which has been touched
by the fingers develops a fingerprint of surface crystallization when strongly heated. The surface has been contaminated
by mineral salts from the fingers. A second striking feature
of the results is the very low growth rate at 24 mol.% Na_2O.
This is very close to the composition of the eutectic between
SiO_2 and the compound $Na_2O.2SiO_2$. When studying the effect of
composition on the ease of glass formation, one frequently finds
that the eutectic compositions are particularly stable and that
the addition of components which reduce the liquidus temperature
increases the stability of the glass. Many examples of this
effect have been given by Rawson (1967).

In the important ternary system $Na_2O-CaO-SiO_2$, the maximum

18

TABLE III

Crystal growth rates in Na_2O-SiO_2 glasses (Dietzel and Wickert, 1956)

Na_2O mol.%	Maximum growth rate $\mu m\ min^{-1}$
0	10
1	620
4	710
8	700
12	640
16	140
20	30
24	1.45
28	10
32	65
36.5	42
39	29
40	115
42	270

growth rate decreases as one approaches the ternary eutectic at 75 w.% SiO_2, 5% CaO, 22% Na_2O (liquidus 725°). Because of its poor chemical durability, a glass with such a high alkali content is of little practical value. The commercial container glass and window glass compositions are closer to the ternary composition 74 w.% SiO_2, 10% CaO, 16% Na_2O, which has a liquidus temperature of about 1000°C and a maximum crystal growth rate of 5 $\mu m\ min^{-1}$. In the manufacture of sheet glass, where a ribbon is drawn continuously from the surface of the melt in a furnace, devitrification problems are frequently encountered due to the existence of pockets of slow-moving glass at temperatures close to the liquidus. For this reason, sheet glass compositions normally contain some MgO in place of some of the CaO. This reduces the liquidus temperature slightly, compared with the MgO-free composition, and reduces the maximum crystal growth rate by a factor which can be as much as 50 (Swift, 1947). Commercial sheet glass compositions also contain some Al_2O_3, which has the effect of increasing the resistance to attack by

atmospheric moisture. A typical sheet glass composition has a
liquidus temperature in the range 930-950°C and a maximum crystal
growth rate in the range 1 to 1.5 μm min^{-1}.

6. Materials which Depend upon Controlled Nucleation and Growth

Materials, the production of which depends upon the controlled growth of a crystalline phase in a glassy matrix, have
been known for many centuries. One example is the ruby-coloured
glass made by adding gold salts to the glass batch. The rapidly
cooled glass contains a high concentration of submicroscopic
nuclei of metallic gold and appears slightly yellow. By subsequent heating, the nuclei can be grown to particles of colloidal
dimensions, when the colour changes to an intense and very beau-
tiful ruby. The colour can be accounted for quantitatively in
terms of the measured optical properties of the metal (Bamford,
1977).

In recent years it has been discovered that the homogeneous
nucleation of gold in a glass can be "catalyzed" by photoelectrons produced by the interaction of short wavelength radiation
with the glass. The photoelectrons are most easily provided by
adding cerium oxide to the glass. In such a glass both Ce^{4+} and
Ce^{3+} ions are present. A light photon of sufficient energy (in
the blue end of the visible spectrum or in the near UV) can eject
an electron from a Ce^{3+} ion, converting it to Ce^{4+}. Neither of
these ions causes any colour in the glass. The photoelectrons
can be captured by gold ions in solution converting them into
gold atoms. If conditions are favourable, these aggregate to
form the nuclei which can be subsequently grown by heat treatment of the glass. It should be obvious that this provides a
means of developing a photographic image in a glass plate, simply by interposing a negative between the radiation source and
the glass (Stookey and Maurer, 1961). An interesting and practically important development of this technique involves the use
of the gold nuclei for the growth of a crystalline silicate phase
which is much more soluble in acid than the parent glass. The
acid-soluble crystals are grown throughout the thickness of the
glass in a pattern determined by the negative used. In this way
extremely complex patterns can be chemically machined in the
glass (Fig.6).

20

Fig.6. Components made from chemically machinable glass.

Practical use is also made of light-irradiation effects in
the photochromic glasses, which contain a high concentration of
submicroscopic crystals of AgCl. Light converts some of the Ag
ions to silver metal, the same process which causes darkening of
photographic film. However, in the photochromic glasses, the
darkening is reversible at room temperature and bleaching may be
considerably accelerated by heating the glass (Smith, 1967).
Returning for a moment to more commonplace materials, white opal
glasses depend for their opacity on the controlled nucleation
and growth of either fluoride or phosphate crystals in a glassy
matrix. These glasses, developed empirically over a considerable period of time, are not easy to make reproducibly. Crystal
growth is markedly affected by small changes in composition.
This in turn affects the opacity and also the strength of the
material.

Finally in this section we come to a very important and
varied class of new materials, the glass-ceramics. Initially a
glass article is made by one of the conventional processes used
in the glass industry. It is then subjected to a two-stage heat
treatment (Fig.7), the first stage being at a relatively low

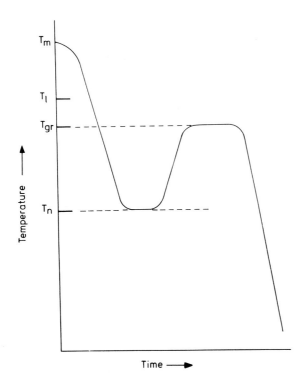

Fig.7.   Temperature-time schedule for making a glass-ceramic.

temperature, T_n, to produce a large number of nuclei. This is
followed by heat treatment at a higher temperature T_{gr} to grow
the crystals to the size required. It is usual for nearly all
the glass to be crystallized.

Generally speaking, the crystal size in the glass-ceramics
is considerably smaller than in conventional ceramics made by
sintering a pre-shaped powder compact. The glass-ceramics as a
class are therefore relatively strong materials; in general the
strength of a polycrystalline ceramic increases as the crystal
size is reduced.

Glass-ceramics having a very low thermal expansion coefficient can be made by choosing compositions in either the Li_2O-
Al_2O_3-SiO_2 system or the MgO-Al_2O_3-SiO_2 system. The low thermal
expansion coefficient is attributable to the major crystalline
phase which is present, i.e. β-spodumene, $Li_2O.Al_2O_3.4SiO_2$ in
the first system and cordierite, $2MgO.2Al_2O_3.5SiO_2$ in the second.
The low expansion coefficient coupled with other very desirable

properties, such as lack of porosity and an ability to accept a
very high surface finish, have led to their application in the
home for top-of-oven ware and electric cooker hob plates. At
the more expensive end of the market they have been used to make
large telescope mirrors, the glass blank being made by casting
the glass into a circular mould.

It is perhaps more than a little ironic that a fundamental
limitation on the very nature of the glassy state, i.e. metasta-
bility and a tendency to devitrify, should have been so well
exploited to produce such a variety of new materials. Their
development represents perhaps the most outstanding series of
innovations which has ever been carried out in the field of glass
technology. The greater part of the credit for this is due to
the laboratories of the Corning Glass Works and to the pioneer-
ing work in those laboratories by S.D. Stookey. An excellent
account of glass-ceramics has been given by McMillan (1964).

E. Immiscibility in Melts and Glasses

Within certain ranges of composition and temperature, some
two-component and multi-component melts are not able to form a
homogeneous single liquid phase. Instead they separate into two
liquid phases differing markedly in composition from each other.
An example is shown in Fig.8 which is the phase diagram of the
system $BaO-B_2O_3$. The region of immiscibility is enclosed by
the dome-shaped dotted line and the horizontal line, below which
crystals of the compound $BaO.4B_2O_3$ are stable in contact with a
B_2O_3-rich liquid phase. Above 1100°C a single-liquid phase is
the most stable arrangement in the B_2O_3-rich half of the system,
i.e. it has the lowest free energy. One of the most interesting
recent scientific developments in the study of glasses has been
the realization that in many well-known glass-forming systems
there exist regions of immiscibility below the liquidus temperature. Figure 9 shows an example in the pseudo-binary system
formed by the compound $Na_2O.4B_2O_3$ and SiO_2 (Rockett et al.,
1965).

It must be emphasised that sub-liquidus regions of immiscibility of this kind are metastable features. At any temperature
below the liquidus temperature, the arrangement of lowest free

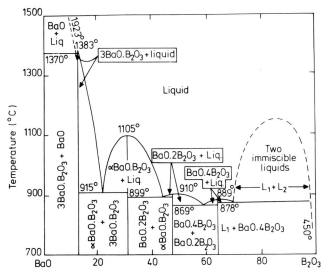

Fig.8. Phase diagram of the system BaO-B₂O₃.

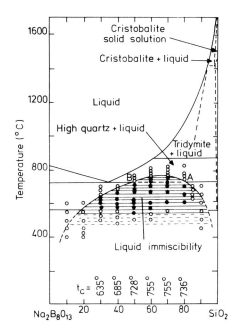

Fig.9. Phase diagram of part of the system Na₂O.4B₂O₃-SiO₂.

energy is one containing one or more crystalline phases. Since
we are dealing with a phenomenon which occurs at relatively low
temperatures, the rate at which the two phases can separate may
be low, especially in glass-forming systems where the viscosity
below the liquidus temperature is high. When the material is
cooled below the transformation temperature no further separation will be possible.

The extensive use of the electron microscope to study glasses
has shown that regions of sub-liquidus, or metastable, immiscibility occur in many systems; even in systems which have been thor-
oughly studied over a period of many years, such as the Na_2O-SiO_2
and Na_2O-CaO-SiO_2 systems. In Fig.10 are examples of the two
types of separation textures which are most frequently observed.

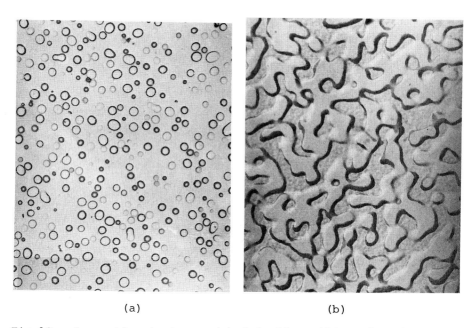

(a)   (b)

Fig.10. Separation textures (a) Soda-lime-silica glass heated
at 740°C for 7½ hr (x14,000) (b) 'Vycor' glass (sodium borosilicate) heated at 700°C for 5½ hr (x24,000).

In Fig.10(a) there are isolated droplets of one phase dispersed in a matrix of the other, whilst in Fig.10(b) both phases
are interconnected as in a sponge saturated with water. The

"sponge" phase is almost pure silica and the "water" phase almost pure sodium borate. If the phase-separated glass is soaked
in dilute acid, the continuous borate phase is dissolved away,
leaving behind a silica skeleton which retains the shape of the
original glass article. If this is now carefully dried and
heated, the skeleton shrinks considerably eliminating all porosity, and one is left with a clear glass containing about 96%
SiO_2. This material, manufactured by Corning Glass Works under
the name "Vycor", is now widely used to make chemical ware. It
was patented as long ago as 1938, many years before the phenomenon of metastable immiscibility had been recognized for what
it is. A good account of "Vycor" and its manufacture has been
given by Volf (1961).

The textures resulting from metastable immiscibility are
often on so small a scale that they cannot be detected even at
the highest magnification obtainable with an optical microscope.
They may not even scatter light to an extent which can be detected by the naked eye. A simple test, which can be useful in
detecting phase separation in what appears to be a homogeneous
single-phase specimen, is to look at the scattering from a laser
beam passing through the glass. If a low-power He-Ne laser is
used, it may be necessary to photograph the specimen using a
relatively long exposure time. Figure 11 shows some photographs
taken in this way. The subject of **metastable** immiscibility has

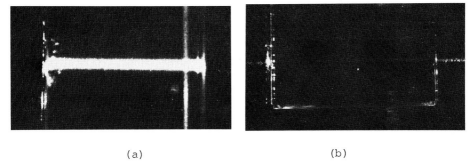

(a)   (b)

Fig.11. Laser light scattering from two specimens of soda-lime-
silica glasses of different compositions (Scott and Rawson,
1973).

been intensively studied in recent years and the literature on
it is extensive. Useful reviews have been made by Charles (1973)
and by James (1975).

The discovery of this type of immiscibility has already had
a marked effect on thinking about the nature of inorganic glasses
and on the approach to the study of the effects of composition
on glass properties. It is clearly important when interpreting
the results of such studies to know whether one is dealing with
single-phase materials or not. Some physical properties, especially viscosity, change markedly when phase separation takes
place and the properties may vary with time in a most remarkable
way as the texture of the material changes during an isothermal
heat treatment (Simmons et al., 1974; Howell et al., 1975;
Simmons, 1977).

F.   The Structure of Oxide Glasses

This is a subject fraught with difficulties. These arise
from the fact that one is dealing with a disordered arrangement,
nearly always consisting of more than one type of atom. In an
X-ray diffraction experiment such a material does not give the
sharp pattern of spots typical of that obtained from a single
crystal and from the positions and intensities of which the
structures of even quite complex inorganic and organic crystals
can be deduced. Instead one obtains a diffraction pattern consisting of two or three rather diffuse concentric rings. The
radial variation in intensity across the pattern can be used to
obtain some structural information, but a detailed structural
determination is impossible. The same difficulties exist when
trying to determine the structure of such "simple" liquids as
molten metals and water. One has to accept this situation, using
the limited information obtainable by X-ray diffraction alone
but looking for other experimental techniques which may be used
to supplement this.

There are now many techniques capable of giving structural
information. Neutron diffraction is one which can give information about the structure as a whole. The ability of atoms to
scatter neutrons differs considerably from their ability to scatter X-rays. By using neutron diffraction, either alone or in

conjunction with X-rays, more detailed structural information
may sometimes be obtained. Spectroscopic techniques are particularly valuable for studying local details of the structure and
the chemical bonding around particular atoms. An excellent account of the use of spectroscopic techniques to study glass
structure has been given by Wong and Angell (1976).

Many glass technologists have attempted either to deduce
structural information or to support particular structural models
by using information on the variation of glass properties with
composition. This approach is full of pitfalls. There are however a few examples, which are best left to later chapters, where
the deductions made from a study of physical properties are almost certainly correct. Very interesting results have been obtained by Westman, Masson and their colleagues who have used
chromatographic techniques to investigate the nature of the anions present in silicate and phosphate glasses containing relatively high percentages of basic oxides (Westman, 1960; Masson,
1977). Finally, in recent years there has been a rapid growth
of activity in the field of "model building". Either by using
pieces of wire and a soldering iron, or by use of the more intellectual activity of computer programming, structures have been
built up, the positions of the atoms have been determined and
the results used to predict the X-ray diffraction pattern of the
material which the modeller is attempting to simulate. This type
of work has led to quite impressive agreement with experimental
results. There seems little doubt that the model approach will
be developed further and will prove increasingly useful as more
powerful computers become available.

Fortunately for the author, this is a book primarily intended
to be a relatively elementary text-book for undergraduate students. It is sufficient for the understanding of the rest of
the text to present only a brief account of what is known about
the structure of oxide glasses. All this information was available as long ago as 1950, if not before, but some detail has been
added since that time.

Prior to the early 1930s, glass technologists tended to think
about glass structure in terms of the physical chemistry of aqueous solutions. This approach was almost entirely speculative
and, on the whole, has not proved helpful. The critical change

in approach which has ever since influenced thinking about the
structure of glasses (especially silicate and borate glasses)
occurred with the publication of a paper by W.H. Zachariasen in
1932. This paper was also speculative in nature and was as much
concerned with proposing a set of structural rules for glass
formation in inorganic systems as with discussing glass structure.
In the author's view, Zachariasen's views on the reasons for glass
formation have not proved helpful - rather the reverse. Here we
shall limit the discussion to Zachariasen's contribution to the
understanding of the structure of silicate glasses. On this point
what he had to say was extremely simple. In effect it was this:
the ordinary rules of crystal chemistry apply to silicate glasses,
just as they apply to silicate crystals. The basic building unit
of a silicate crystal is a SiO_4 tetrahedron. These tetrahedra
are joined together at their corners only, not along their edges
or faces. Exactly the same situation exists in a silicate glass.
In the crystal the same geometric pattern is repeated over large
distances throughout the material. Angles between bonds of the
same type are constant and distances between pairs of atoms such
as the silicon and oxygen atoms are constant. However, in the
glass, the bond angles and bond distances vary.

It is difficult to illustrate Zachariasen's ideas in three
dimensions, but very easy to do so in two. Figure 12, which is
probably the most frequently reproduced diagram in glass technology, illustrates the structure of a hypothetical two-dimen-
sional compound X_2O_3. The diagram shows both the crystalline
and the glassy form of the material. In a series of papers,
published in the 1930s and early 1940s, B.E. Warren and his colleagues confirmed Zachariasen's model by X-ray diffraction studies
of a number of oxide glasses of simple composition. More recently
Warren and Mozzi (1969) have examined the structure of glasses
using improved techniques now available. Although the recent work
leads to conclusions a little different from the earlier work,
the differences are slight and need not concern us here.

One feature of the glassy and crystalline structures considered by Zachariasen is that, since they are built up from
similar polyhedra joined only at their corners, the structures
are relatively open and contain relatively large voids. Warren's
X-ray studies amplified Zachariasen's picture by indicating where

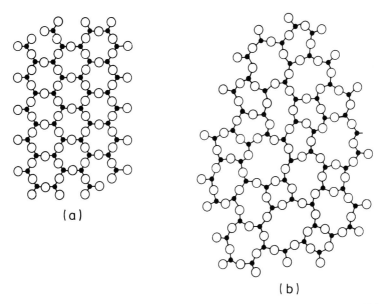

Fig.12. Structure of the compound X_2O_3 in (a) the crystalline and (b) the glassy form.

the sodium ions in a Na_2O-SiO_2 glass are situated. This is illustrated two-dimensionally in Fig.13. As may be seen, the so-

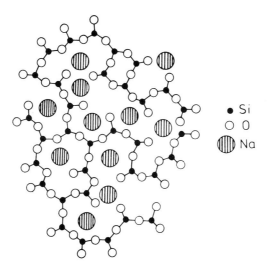

Fig.13. Structure of a Na_2O-SiO_2 glass.

dium ions are situated in the voids in the network formed by the SiO_4 tetrahedra. Note also that the oxygen ions introduced together with the sodium ions in the form of Na_2O are incorporated by the rupture of some of the Si-O-Si linkages which are present in the SiO_2 glass. It is generally agreed that the structure in the vicinity of these so-called non-bridging oxygen ions carries a negative charge to neutralize the positive charge of the sodium ions and that these sodium ions are likely to be found adjacent to the singly-bonded oxygen ions so that electro-neutrality is maintained locally in the structure. The Na-O bonds are largely ionic in character and are much weaker than the Si-O bonds, which have a strongly covalent character and hence are more directional. It is believed that most of the oxides of Group I and Group II elements enter the glass structure in a similar way to Na_2O.

An addition of Al_2O_3 may be incorporated in another way. It is well-known that Al^{3+} can replace Si^{4+} in the structure of crystalline silicates, thus forming AlO_4 tetrahedra. This, however, can only occur if other positively charged atoms are introduced at the same time, i.e. for each Al^{3+} ion replacing Si^{4+} in the tetrahedral position, one must also introduce a single Na^+ ion so that the sum of positive charges is unchanged. One may therefore expect that the compound $Na_2O.Al_2O_3.6SiO_2$ (which forms a glass) will have a structure very similar to that of silica glass itself. All the aluminium ions may be incorporated in the tetrahedral position, the equal number of sodium ions finding themselves in voids - probably in the neighbourhood of the AlO_4 tetrahedra. This structure need have *no* so-called "non-bridging" oxygens. However, it is also known that in crystalline silicates the aluminium ion may be surrounded octahedrally by oxygen ions, in which case obviously it is not replacing silicon ions in the network. This is likely to happen if an insufficient number of sodium ions is introduced at the same time to obtain the balance of charges referred to above.

This is probably sufficient as an elementary introduction to the problem of structure, provided that it is realised that the picture presented is likely to be correct in broad outline only. In reality the situation is not likely to be so simple. For example, it is well-known that most samples of silica glass are oxygen-deficient. This gives rise to defects in the structure. Similar defects *may* exist in binary and ternary oxide glasses.

There is also some evidence to suggest that the alkali ions may not be uniformly distributed throughout the structure. Even in an apparently homogeneous glass they may tend to form clusters. Also, there are certain special structural features about certain important types of oxide glasses, especially the borate glasses, which need to be mentioned. They will be referred to later.

It must not be imagined that the Zachariasen "random network hypothesis", as it has been called, can be applied to all inorganic glass-forming materials. It is of no value when considering the structure of vitreous selenium and it is very unlikely that it has very much to say about the chalcogenide glasses and the metallic glasses. However, for the commercially important silicate glasses, it has provided a very useful mental picture for our structural ideas.

Suggestions for Further Reading

1. Turnbull, D. (1969) Contemp. Phys., 10, 473-88.
2. Jones, G.O. (1971) "Glass", 2nd Edn., revised by S. Parke. Chapman and Hall - Science paperbacks, London. Chapters 1 to 4.
3. Pye, L.D., Stevens, H.J. and La Course, W.C. (1972) "Introduction to Glass Science". Plenum, New York, 722 pp. Chapter 1.
4. Doremus, R.H. (1973). "Glass Science". Wiley-Interscience, New York. 349 pp. Chapters 1 to 7.
5. Wong, J. and Angell, C.A. (1976) "Glass Structure by Spectroscopy". Dekker, New York. 864 pp. Chapter 1.

32

CHAPTER II

VISCOSITY

The way in which the viscosity of a glass melt varies with temperature is the most important factor in determining the kind of shaping processes which can be used in making glass articles. Any account of glass fabrication processes should begin with a discussion of the viscosity of glass melts and its variation with temperature. In the continuous automatic processes used in the glass industry, the forming machines must be fed with glass of constant viscosity, otherwise dimensional variations and other defects will be encountered. The glass technologist must therefore have a good understanding of the factors which affect the viscosity of the melt.

At the maximum temperature (ca. 1550°C) in a glass melting furnace producing a container glass or flat glass composition, the melt has a viscosity of the same order as that of golden syrup at room temperature. At 500°C the same glass has a viscosity which is so high that the material behaves, under most conditions, as an elastic solid. In the temperature range between 1500° and 500°C, the viscosity rises steadily as the temperature falls. There is no sudden discontinuity.

Almost all glass articles are made whilst the material is cooling. Throughout the process the glass surface is consequently at a lower temperature than the interior and hence has a higher viscosity. Thus a second factor which is of primary importance in all glass fabrication processes is the heat transfer between the glass and its surroundings, since the rate of heat transfer determines the magnitude of the temperature gradients.

A. Definition

The shear viscosity, or coefficient of shear viscosity (usually referred to simply as the viscosity), is most easily defined in terms of the flow situation shown in Fig. 14. The liquid is confined between two parallel plates, the lower one of which is stationary whilst the upper moves at a constant ve-

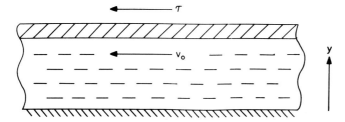

Fig. 14. Definition of viscosity.

locity v_o, the distance between the plates remaining constant.
For a very large number of liquids, the velocity in the liquid
varies linearly with distance between the plates. At the surface of the lower plate the liquid velocity is zero and at the
surface of the upper plate it is v_o. To maintain this motion,
it is necessary to apply a force to the upper plate in the direction of its motion in order to overcome the "forces of friction" within the liquid. The force per unit area applied to
the upper plate, and, through it, to the upper surface of the
liquid layer, is proportional to the velocity gradient in the
liquid, dv/dy, and the proportionality constant is a property
of the liquid, its viscosity η. Thus

$$\tau = \eta.dv/dy \qquad (7)$$

This simple laminar flow situation exists only when v_O is
below a certain value. At high velocities, turbulent flow develops. However, turbulent flow is never encountered in glass
melts because of their high viscosity.

Also we should note that for some liquids the variation of
velocity with position in the flow of Fig. 14 would be non-lin-
ear. Such liquids are said to be non-Newtonian. Fortunately,
any deviation of glass melts from the ideal Newtonian behaviour
is small enough to be neglected in the great majority of situations of practical interest.

In the S.I. system of units, τ in equation (7) is in N m^{-2},
or Pascals (Pa) and the unit of dv/dy is s^{-1}: hence the units
of η are N s m^{-2} or Pa s. In the c.g.s. system, τ is in dyne
cm^{-2} and the units of η are dyne s cm^{-2}. This unit is called
the "poise" and has been used almost universally in the glass
literature until the present day. It is likely to continue in
use in the glass industry for many years to come. Fortunately

the conversion from one system to the other is easy: 1 poise = 0.1 Pa s.

It is very useful in a number of contexts to recognize the close analogy which exists between the equations relating stresses and velocity gradients in a viscous liquid on the one hand, and those relating stresses and displacement gradients (or strains) in an elastic solid on the other. The analogy can be fully appreciated only by considering the equations in detail, which would not be appropriate here. However, it is easy to see the close analogy between the definition of shear viscosity as given above, and the definition of the shear modulus, G, of an elastic solid. In elementary texts, the shear modulus is defined in terms of the distortion of a rectangular block subjected to a shearing stress (Fig. 15). The relationship between

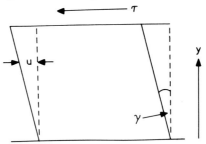

Fig. 15. Definition of shear modulus.

the applied shearing stress τ, and the angle of shear γ, is $\tau = G\gamma$. If γ is small, $\gamma = du/dy$ and

$$\tau = G.du/dy \qquad (8)$$

where u is the displacement of a point in the direction of the applied shearing stress. This has the same form as equation (7), which defines the shear viscosity. For reasons that will become apparent later, it is appropriate to note here that for an incompressible elastic solid, i.e. one for which the compressibility modulus K is infinitely high, E = 3G, where E is the Young's Modulus of Elasticity. In problems involving the flow of viscous liquids, the main effect of the applied stress is to produce velocity gradients in the material, the change in volume due to the applied stresses being negligible. Thus no serious error results from treating the viscous liquid as an incompressible material. We shall see later a number of examples in which useful results can be obtained by taking well-known equations for the deformation of elastic solids under load and replacing in these equations G or $E/3$ by η to obtain corresponding equations for the <u>rate</u> of deformation of a viscous liquid under load.

B. Viscosity values

The viscosities of a number of Newtonian liquids are given in Table IV whilst Fig. 16 shows the viscosity-temperature curves

TABLE IV

Viscosities of some simple liquids

	Temperature °C	Viscosity Pa s
Water	20	1.002×10^{-3}
Sodium chloride	841	1.3×10^{-3}
Iron	1400	2.25×10^{-3}
Glycerol	- 42	6.71×10^{-3}
	20	1.49
Aqueous sucrose solution (75%)	20	2.33

of two commercial glass compositions. In the melting furnace the viscosity of the melt is in the range 10 - 100 Pa s. At the start of most shaping processes the glass has a viscosity of about 10^3 Pa s. During the process, forces are applied to the glass and heat is extracted from it. The process must be operated in such a way that, when the glass product is released from the machine, the viscosity of the surface layers must be high enough that the glass does not deform under its own weight. The glass product is subsequently "annealed", i.e. it is subjected to a controlled temperature-time schedule which first brings the glass to a substantially uniform temperature and then cools it in such a way that the stresses remaining in the glass at room temperature are below a specified value. As will be seen later, the specification of the annealing temperature schedule requires information on the variation of the viscosity of the material with temperature.

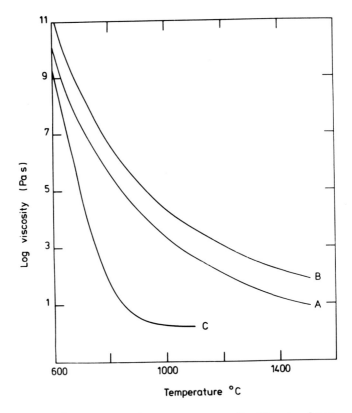

Fig. 16. Viscosity-temperature curves. A. Sheet glass.
B. Borosilicate glass (Physical properties Committee, 1956).
C. Experimental borosilicate glass: 35 w% SiO_2, 40% B_2O_3,
25% Na_2O.

Curve C in Fig. 16 is for a borosilicate glass melt of high
B_2O_3 content. It would be difficult to produce shapes from this
composition by many of the processes commonly used in the glass
industry, e.g. blowing of hollow articles in a metal mould, drawing a sheet from the free surface of a melt, or fusing together
parts in a flame. At high temperatures the viscosity of the melt
is low and varies little with temperature. However, at lower
temperatures there is a very rapid rise of viscosity with falling
temperature. This glass would set very quickly as it cooled and
it would be difficult to control the shaping process. This does
not necessarily mean that such a glass would be of no practical
interest. Some glasses which have valuable optical and electrical properties have similar, or even less favourable, viscosity-

temperature curves and are therefore not suitable for shaping by
the processes referred to. However, shapes can be made by casting blocks into a mould, annealing, and then producing the shape
required from the cold block by mechanical methods, e.g. grinding, cutting and drilling. Also, some glasses, e.g. the so-
called "solder glasses" (which are used for sealing together components in colour television tubes, for example) and vitreous
enamels are required only in the form of powders or granules.
They are subsequently fused to form coatings on a solid substrate. The considerations which are relevant when determining
the most desirable viscosity-temperature curve for materials of
these types are quite different from those which apply in the
manufacture of hollow glassware or flat glass.

C.   Analysis of Some Simple Problems in Viscous Laminar Flow

It was pointed out in the introduction to this chapter that
two factors of great importance in the control of glass manufacturing processes are the viscosity-temperature relationship of
the glass being processed and the heat transfer between the glass
and its surroundings.

In the present section, consideration will be given to two
situations which illustrate the interplay of these two factors.
At the same time they illustrate how equation (7) can be integrated to give formulae for the velocity in the glass as a function of position.

1.   Flow of Glass Over a Horizontal Plane

The first problem to be considered relates to the flow of
glass in that part of a glass-melting furnace in which the glass
is being brought to the right temperature and hence to the right
viscosity for feeding to the glass-forming machine. In this
section of the furnace, the glass is being fed as a relatively
shallow layer along a channel made from refractory bricks. This
channel is several metres long and provision is made to apply
controlled heating and cooling to the glass as it passes along
the channel so that, at the end, the glass temperature has the

required value and temperature gradients within the glass have
been reduced to a minimum. The width of the channel depends
upon the type of glassware being manufactured. In a container
glass furnace, the width is usually less than 1 metre, but in a
factory making flat glass the channel is somewhat wider than the
glass ribbon which is being manufactured, i.e. several metres.

In a simple analysis it is not possible to consider the
effect of the vertical side walls of the channel on the velocity
distribution in the glass. We must limit ourselves to an analysis of the flow over a horizontal plane which is assumed to be
of infinite extent in the horizontal direction at right angles
to the direction of flow (Fig. 17). The driving force which

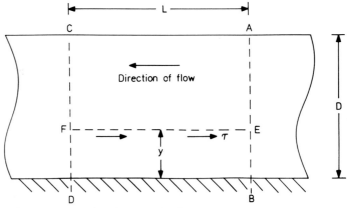

Fig. 17. Flow over a horizontal plane.

causes the glass to flow from right to left in the diagram is a
pressure gradient in the glass arising from the fact that raw
material is being fed continuously into the melting section of
the furnace, which is away to the right of the diagram, and is
being withdrawn continuously at the same rate to feed the forming machine, which is off to the left. The hydrostatic pressure,
p_A, at the glass surface on plane AB is thus greater than the
pressure, p_C, at the surface on plane CD. Both these pressures
increase with depth in the glass, but do so at the same rate.
Thus, for any pair of points such as E and F which are at the
same distance from the surface, the pressure difference is the
same as at the surface.

$$p_E - p_F = p_A - p_C .$$

We consider the balance of forces acting on a rectangular block

of glass AEFC of unit thickness measured in a direction normal
to the plane of the paper. The pressure difference between
planes AB and CD gives rise to a net force acting towards the
left of magnitude $(p_A-p_C)(D-y)$, where D is the depth of glass
in the channel. Since we are limiting the analysis to the steady
state situation where at any point the glass velocity does not
vary with time, it must follow that this net force due to the
pressure gradient must be balanced by an equal and opposite force.
This is the force acting on plane EF arising from the viscous
shearing stress τ on that plane. The stress τ gives rise to a
force τL where L is the distance between the two planes AB and
CD. Hence

$$\tau L = (p_A-p_C)(D-y) \ .$$

Combining this with equation (7) we obtain

$$dv/dy = (p_A-p_C)(D-y)/\eta L \ . \tag{9}$$

If there are no temperature gradients in the glass, η is constant and this equation can be integrated to give

$$v = (p_A-p_C)(Dy-y^2/2)/\eta L \ . \tag{10}$$

Note that this equation is obtained by using as a boundary condition the fact that v=0 at the interface with the refractory,
i.e. at y=0. This no-slip condition always obtains at an interface between a solid surface and a liquid which wets it. The
equation shows that the velocity increases with y to a maximum
value at the surface (y=D). The preceding equation for the velocity gradient, dv/dy, shows that this is zero at the surface.
This must be so because the surface is an interface between the
glass and a gas phase of negligible viscosity. Such an interface must be free from shearing stresses.

If the temperature, and hence the viscosity, of the glass
varies with the distance y, the differential equation cannot be
integrated analytically. However, it can be integrated numeri-
cally using the equation

$$v = (p_A-p_C) \cdot L^{-1} \int_o^y (D-y) \cdot \eta_y^{-1} \cdot dy \tag{11}$$

To carry out the integration one must have data for the variation of viscosity with y. In practice this amounts to knowing
the vertical temperature distribution in the glass and the variation of glass viscosity with temperature. Note that whatever
the vertical temperature may be, the result of the numerical in-

tegration must lead to a velocity distribution in which the velocity gradient dv/dy is zero at the glass surface. Figure 18 shows
two velocity distributions, one which is obtained if η is constant, the other being a possible velocity distribution for a
situation in which η varies with y.

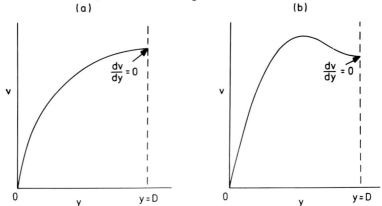

Fig. 18. Velocity distributions in glass flowing over a horizontal plane.

In order to analyse more realistic problems than this one,
e.g. the flow of glass in a rectangular channel and the flow in
the melting furnace itself, it is necessary to go into fluid mechanics in some detail. Solutions of the relevant equations can
only be obtained using a computer. Much work is now being done
in this area and the interested reader is referred to review articles for further information (Rawson 1974, 1977).

2. Heat Transfer from Glass Inside a Cylindrical Mould

In the second example, we shall be concerned with flow in
glass after temperature gradients have been produced in it by
heat transfer between a cylindrical charge of hot glass and the
internal surface of a metal mould, the temperature of which is
considerably less than that of the glass. This situation is encountered in the manufacture of glass containers, the glass charge
being fed into the mould at a temperature of approximately 1100°C
whilst the internal mould surface is initially at a temperature
in the range 400° - 450°C.

Trier (1961) carried out experiments in which he measured
the flow in the glass (in response to pressure forces) after it

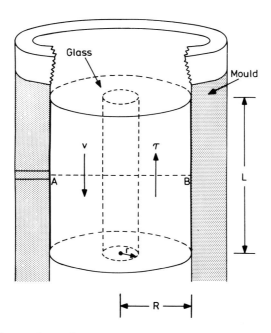

Fig. 19. Trier's method for measuring temperature distributions produced in glass by cooling in a mould.

had been in contact with the mould for a controlled period of time. The experiment is shown schematically in Fig. 19. This shows the glass charge in the mould. Immediately after the hot glass enters the mould, a steel needle is driven through it along the diameter line AB. The needle is then immediately retracted leaving behind it a line of bubbles. After a time t, which was varied in the experiments from 2 to 15 seconds, a vacuum is applied to the space below the glass, which up to this point has been supported on a platform consisting of a nest of steel cylinders. The difference between the atmospheric pressure acting on the upper face of the glass and the vacuum under the lower face causes the glass to flow in a vertical direction. When the glass is cold, the cylindrical block is removed from the mould and sliced into two equal parts along the diametral plane containing the bubble trace. Figure 20 is a photograph showing one of the results. The amount of axial movement decreases towards the wall. Near the wall where the glass has been chilled by contact with the mould, the amount of axial flow in the glass is very small. It is greatest at the centre of the charge. From

Fig. 20. Flow trace (Trier, 1961).

measurements of the amount of axial movement, Trier was able to calculate the viscosity distribution in the glass at the instant when the vacuum was applied and hence obtain the temperature distribution, which was the purpose of the experiment.

To obtain equations which allow one to determine the viscosity of the glass as a function of r, the radial distance from the vertical axis of the glass cylinder, we again make use of the principle of equilibrium of forces. The downward-acting force due to the pressure of the atmosphere acting on the circular sec- of the upper surface or radius r is $p_A \pi r^2$. The upward-acting force on the lower surface of the cylinder of radius r is, of course, zero. This force difference is balanced by forces due to the shearing stress, τ, acting over the cylindrical surface of radius r and length L, i.e. by the force $\tau \, 2\pi r L$. Hence

$$\tau \, 2\pi r L = p_A \, \pi r^2 ,$$

and

$$\tau = \eta_r \, dv/dr ,$$

where η_r is the viscosity of the glass at the distance r from the central axis. Hence

$$\eta_r \, dv/dr = p_A \, r/2L . \tag{12}$$

From measurements on the bubble trace, dv/dr may be obtained at any point and η_r may then be calculated from the above equation.

D. Measurement of Viscosity

The range of viscosity which is of interest in the manufacture and use of glasses is very wide (1 to 10^{13} Pa s) and it is not possible to make measurements over the whole of this range by any one method. It is convenient to describe the methods used by dividing them into two groups, one group being used for measuring relatively low viscosities, i.e. at high temperatures, and the other for measuring high viscosities. It is not possible to give a viscosity value defining the boundary between these two groups since the viscosity range over which any one method is capable of giving accurate results depends to some extent on details of apparatus design.

In measuring the viscosity of glasses, especially in the region of high viscosities where the viscosity varies rapidly with temperature, it is important to pay careful attention to the measurement of temperature. The thermocouple used for measuring the glass temperature should be placed as close to the glass as possible and ideally it should be in thermal contact with it - if this will not interfere with the viscosity measurement. The viscometer furnace must be carefully designed so that the glass is in a zone of constant temperature and an adequate time should be given at each temperature of measurement to ensure that the glass has reached an equilibrium temperature before the measurement is made. It is also desirable to maintain a stock of a standard glass of homogeneous composition, the viscosity of which can be measured from time to time to ensure that no change has taken place in the apparatus used. This is good practice when making measurements of any physical property if the measurements are likely to be carried out over a considerable period of time. It is especially important to do so when one is concerned with property measurements at high temperatures. Changes in furnace characteristics and thermocouple calibrations, which seriously affect the accuracy and repeatability of the measurements, can go undetected all too easily.

1. High Temperature Measurements

    The most widely used method for measuring the viscosity of
glasses at high temperatures is the rotating cylinder method.
Measurement over the range from 10 to 10^5 Pa s is possible with
most designs of apparatus of this type, but a wider range than
this can be covered by incorporating special features in the design which permit the apparatus to be used in a number of different ways. Descriptions of the apparatus and measurement
techniques by Dietzel and Brückner (1955) and by Napolitano et
al. (1965) are particularly helpful for the accounts they give
of how to extend the range of measurement.

    Figure 21 illustrates some features of the Dietzel and
Brückner apparatus and this will be used to explain the principles of the method. Cylinders A and B are made from a platinum-

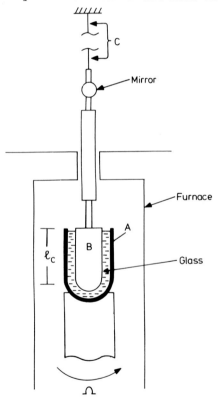

Fig. 21. Rotating cylinder viscometer (Dietzel and Brückner).

rhodium alloy, B being suspended co-axially in A from a torsion wire C. The space between the two cylinders is filled with the glass, the viscosity of which is to be measured. The thermo-couple used to measure the glass temperature can be mounted in an axial hole drilled in the inner cylinder. The usual method of using a viscometer of this kind is to rotate the outer cyl-inder at a constant speed of Ω radian s^{-1} and to determine the torque T so produced on the inner cylinder by measuring the an-gle of twist θ at the lower end of the torsion wire. Using the apparatus in this way, viscosities may be measured over the range from 3 to 6 x 10^4 Pa s. To measure lower viscosities, down to 0.1 Pa s, cylindrical inertia weights are mounted on the suspen-sion at the lower end of the torsion wire. Cylinder A is held stationary and, after turning B through a small angle, it is re-leased. From the damping of the torsional oscillations of cyl-inder B the viscosity of the glass can be calculated. The range can be extended to higher viscosities (10^7 Pa s) by observing the slow return of the central cylinder to its original position after the outer cylinder has been rapidly turned through a small angle and then held fixed (the aperiodic method). Napolitano et al.(loc. cit.) were able to use their apparatus over an even wider range (1 to 10^9 Pa s).

The determination of the viscosity-temperature curve of a glass over the full measurement range of a rotating cylinder viscometer is a lengthy operation. A significant reduction in the time required can be achieved by using a design of apparatus in which the outer cylinder also serves as the heating element (Tiede, 1959). It is heated by passing a heavy current through the platinum cylinder. By this method the glass can be brought quickly to the temperature at which the next measurement is to be made.

The derivation of the equation which applies to the "normal" mode of operation is given in a number of elementary textbooks. This elementary theory ignores end effects. It cannot take into account the effect of the flow in the region between the lower ends of the two cylinders, i.e. it treats the problem as if both cylinders were of infinite length. Although the cylinder length, l_e, enters into the equation, this is an "effective length" which is usually determined by experiment using a liquid of known vis-cosity.

In the region remote from the ends, any point in the glass follows a circular path. For a point distant r from the centre, the angular velocity is ω and the tangential velocity v_θ is then $v_\theta = \omega r$. Differentiating this equation with respect to r gives

$$dv/dr = r\, d\omega/dr + \omega .$$

The second term on the right hand side represents a rigid body rotation in the glass about the axis and is not relevant so far as shear flow in the glass is concerned. The shear stress τ_θ at r is then simply

$$\tau_\theta = \eta\, r\, d\omega/dr .$$

The torque T which causes the twist in the suspension is

$$T = 2\pi r l_e \tau_\theta r ,$$

and this is constant within the glass in the space between the two cylinders.

Combining the two previous equations, a differential equation is obtained, which on integration gives the angular velocity Ω at the outer cylinder in terms of the dimensions of the apparatus and the viscosity of the glass. Re-arrangement of this equation gives the equation used to calculate the viscosity

$$\eta = (T/4\pi\Omega l_e).((r_o{}^2 - r_i{}^2)/(r_o{}^2 r_o{}^2)) \qquad (13)$$

where r_i and r_o are respectively the external radius of the inner cylinder and the internal radius of the outer cylinder.

As stated earlier, the effective length, l_e, is usually determined by calibration. However, for the Dietzel and Brückner apparatus, an exact analysis can be carried out of the flow in the region of the hemi-spherical ends of the cylinders. This shows that for the radius ratio r_o/r_i of 2.48 used in their design

$$l_e = l_c + 2r_i/3 ,$$

where l_c is the length of the cylindrical section of the two cylinders.

When the instrument is being used for measuring low viscosities by the torsional damping method, one determines, from the decay in the amplitude of the oscillations, the logarithmic decrement λ defined by

$$\lambda = \ln(\theta_n/\theta_{n+1}) ,$$

where θ_n and θ_{n+1} are successive amplitudes of the angular deflection, each measured on the same side of the point of zero

deflection. The viscosity is then calculated from a formula derived from the equation of motion of the inner cylinder

$$\eta = (2/A).(CI/(1 + 4\pi^2/\lambda^2))^{\frac{1}{2}}. \tag{14}$$

For the aperiodic method, the equation is

$$\eta = (C/A).(t_2-t_1)/\ln(\theta_2/\theta_1). \tag{15}$$

In these equations, A is a dimensional constant of the apparatus
given by

$$A = 4\pi l_e r_1^2 r_0^2/(r_0^2-r_1^2). \tag{16}$$

C is the torsional stiffness of the suspension, i.e. the constant
which relates the torque in the suspension to the angular deflection θ by

$$T = C\theta ,$$

and, in equation (15), θ_1 and θ_2 are angular deflections measured
at the times t_1 and t_2 respectively. I in equation (14) is the
moment of inertia of the assembly suspended from the wire.

Another method which has been used extensively for viscosity
measurements at the lower end of the range of interest is the
counterbalanced sphere method. This is one of a number of methods based on the Stokes equation, which gives the viscous drag
force, F, acting on a sphere of radius r moving with a velocity
v in a liquid of viscosity η

$$F = 6\pi r v \eta . \tag{17}$$

A platinum alloy sphere is suspended in the glass from the left
hand arm of a modified chemical balance. If the weight on the
right hand pan of the balance W is greater or less than that required for balance, the sphere will move up or down at a velocity, v, which can be measured. The viscosity of the glass can
be calculated from the slope of the straight line obtained by
plotting v against W. A correction factor has to be applied to
take account of the fact that the Stokes equation is exact only
if the diameter of the container is very large compared with that
of the sphere. This factor may be obtained either by calibration with a liquid of known viscosity or by using one of a number of correction formulae which have been obtained to take account of the effect of finite container size. An apparatus of
this type described by Shartsis and Spinner (1951) gives satisfactory results over the viscosity range from 0.5 to 10^3 Pa s.

Having just stated the Stokes equation, it is worth drawing
attention here to its application in calculating the rate of rise

of gas bubbles in a glass melt. Bubbles are produced by chemical
reactions between the constituents of the glass batch or by air
being trapped as the batch powder melts. The great majority of
these bubbles must be removed in the melting process, otherwise
the glass will not be of acceptable quality. The buoyancy force
acting upwards on a bubble of radius r in a glass melt of density
ρ is $4\pi r^3 \rho g/3$. When the bubble is moving upwards at a constant
velocity, the buoyancy force is exactly balanced by the viscous
drag force given by the Stokes equation. Hence the rate of rise
of the bubble is

$$v = 2r^2\rho g/9\eta \ . \tag{18}$$

For a bubble of radius 10 μm in a melt of density 2.3 x 10^3 Kg
m^{-3} and viscosity 10 Pa s, v = 4.7 x 10^{-3} m s^{-1}. It is easy to
appreciate from this result why glass melting is a slow process.
However, it should be pointed out that very small bubbles can be
eliminated by the gas in the bubble dissolving in the melt.

2. Low Temperature Methods

The most widely used method for measuring glass viscosities
at relatively low temperatures involves measuring the rate of
extension of a fibre subjected to an axial stress. Good accounts
of this method have been given by Lillie (1931, 1954) and by Boow
and Turner (1942). The fibre of carefully measured dimensions,
usually 100 mm long and 1 mm in diameter, is suspended vertically in an electric tube furnace. A metal liner inside the furnace tube helps to maintain a uniform temperature along the
length of the fibre. The fibre is stressed by weights added to
a scale pan suspended from the lower end of the fibre. The rate
of extension is measured either by using an optical lever system,
or by a suitable electrical transducer.

The formula used to calculate the viscosity may be obtained
by making use of the analogy, referred to earlier, between the
equations relating stresses and strains in an incompressible
elastic material and those relating stresses and strains in an
incompressible viscous liquid.

The axial strain e_z produced by an axial stress σ_z in an
incompressible elastic solid is

$$e_z = \Delta l_0/l_0 = \sigma_z/E = \sigma_z/3G \ .$$

l_O is the initial length of the fibre and Δl_O is the extension produced by the stress. The corresponding equation for the rate of extension dl_O/dt of a fibre made from an incompressible viscous material is

$$l_O^{-1}.dl_O/dt = \sigma_z/3\eta \ , \tag{19}$$

from which the viscosity may be calculated.

Some glasses are not easily drawn into fibres and, for these, other methods for measuring viscosity may be used. One is to prepare a rectangular bar by cutting and grinding from a cast block of the glass. The bar is mounted in a furnace and is supported horizontally by knife edges at each end. A vertical load F is applied via a third knife edge midway between the supports and the rate of sag, ds/dt, at the mid-point is measured. This bending beam method has been described by Hagy (1963, 1968).

The formula derived from elementary bending beam theory for the deflection s at the centre of a beam made from an elastic solid and loaded in the way described is

$$s = FL^3/4Ebd^3 \ .$$

The corresponding equation for the rate of sag of a beam made from a viscous glass is

$$ds/dt = FL^3/12\eta bd^3 \tag{20}$$

where L is the distance between the supporting knife edges, b is the width and d the thickness of the beam.

Other methods which are suitable for glasses which cannot be drawn into fibres are the parallel plate method described by Hagy (1963) and by Fontana (1970), and the indentation method described by Douglas et al. (1965). The parallel plate method uses a circular disc of glass which is sandwiched between metal plates. The change in the distance between the plates is measured when a force is applied, compressing the disc in the direction normal to the plate surfaces. The shape of specimen used in the penetration method is not very critical, provided that one face is flat and the specimen can be rigidly supported in the furnace so that the flat surface is uppermost and horizontal. A normally acting force is applied to a steel ball bearing which is in contact with the specimen surface. The rate at which the ball penetrates into the glass is measured. As before, the formulae used for calculating the viscosity may be obtained by analogy from corresponding formulae derived for elastic solids loaded in the same way.

The viscosity range covered by the extension of a fibre method and the penetration method is approximately 10^9-10^{12} Pa s, by the bending beam method 10^7-10^{13} Pa s and by the parallel plate method 10^4-10^8 Pa s.

3. Viscosity Fixed Points

The methods for measuring viscosity described in the previous section are somewhat time consuming. For routine control purposes in the glass industry greater use is made of standard tests which are relatively quickly carried out and which involve measuring the temperature at which a vertically suspended fibre of specified dimensions extends at a specified rate. During the test the fibre is heated at a constant controlled rate and the rate of extension is measured continuously whilst the fibre is heated.

In one such test, specified in ASTM Standard C338-57, a fibre 235 mm long and 0.55-0.75 mm in diameter is suspended in a furnace of specified design in such a way that only the upper half of the fibre is heated. The rate of heating is controlled to lie between 4 and 6° min^{-1}. The temperature at which the rate of extension of the fibre is 1 mm min^{-1} is defined as the ASTM or "Littleton" softening point of the glass. The viscosity of the glass at the softening point is approximately $10^{6 \cdot 6}$ Pa s, the exact value depending to some extent on the density of the glass. This is one of three so-called "viscosity fixed points" commonly quoted in commercial catalogues giving glass properties.

Another ASTM specification (C336-64T) specifies the apparatus and method of test for determining the "annealing" and "strain" points. The fibre is loaded with a 1 Kg weight. Again, only the upper part of the fibre is heated. The rate of extension is measured as the fibre cools (Lillie, 1931, 1954). The annealing point is determined on a fibre of approximately 0.65 mm diameter. The cooling rate is approximately 4°C min^{-1} and the annealing point is that temperature at which the rate of elongation is 0.135 mm min^{-1}. The effective fibre length is 230 mm, i.e. that part heated by the furnace. The strain point

is the temperature corresponding to a rate of elongation of
0.0043 mm min^{-1} when measured by this method and is usually
determined by extrapolation from the annealing point results.
According to Lillie (1954), the viscosities corresponding to the
annealing and strain points are 10^{12} and $10^{13\cdot5}$ Pa s respectively.

<h3>E.   Variation of Viscosity with Temperature</h3>

<h3>1.   Activation Energy for Viscous Flow</h3>

Viscous flow in a liquid involves the movement of atoms or
groups of atoms relative to one another. It is not possible to
say in detail what type of movement is involved and there is
every reason to believe that this will differ from one type of
liquid to another. However, whatever the detail, the movement
of the flow unit is resisted by adjacent atoms or molecules.
The flow unit has to overcome an energy barrier, the energy required being supplied by the thermal energy at the prevailing
temperature. As the temperature rises there is an increase in
the probability of a particular unit possessing sufficient energy to surmount the energy barrier. Considerations such as
this lead to equations of the form

$$\eta = A \exp(E_\eta/RT) \ , \tag{21}$$

or

$$\eta = BT \exp(E_\eta/RT) \ , \tag{22}$$

where A and B are constants for the glass, T is the temperature
in °K and R is the gas constant. E_η is a measure of the magnitude of the energy barrier and is usually called the "activation energy for viscous flow".

Equations of this form are familiar in discussions of many
transport phenomena in solids and liquids, e.g. diffusion and
electrical conduction. However, it is clear that the simple
formulae given are by no means satisfactory even within one
class of liquids, for example that of the oxide glass-forming
melts. A plot of ln η against the reciprocal of the absolute
temperature should give a straight line, from the slope of which
E_η can be calculated. Figure 22 shows that although for molten
SiO_2 and GeO_2 a straight line is indeed obtained over a very
wide temperature range, the results for other melts show a very

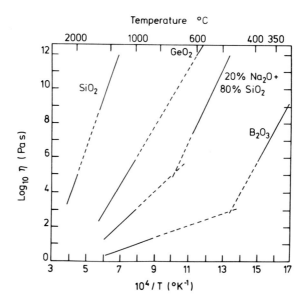

Fig. 22. Viscosity data for some oxide melts.

marked deviation from linearity, the value of E_η being much greater in the low temperature region than at high temperatures. For a soda-lime-silica glass, similar in composition to container glass, the activation energy increases from 55 Kcal mole^{-1} at 1500° to 115 Kcal mole^{-1} at 550°C.

One factor which may account for at least part of the increase is a change in liquid structure with temperature, the structure becoming more open as the temperature rises. The interpretation of the viscosity-temperature relationship in liquids is a subject which has attracted considerable interest and the literature on it is extensive (see for example Macedo and Litovitz, 1965 and Barton, 1971). However, it is probably fair to say that, as yet, no single theory has emerged which is generally accepted as giving a satisfactory account of the viscosity of glassforming melts. This is a state of affairs which need not be regarded either with disappointment or surprise. Because of the inherent difficulty of obtaining structural information, our understanding of liquids and their properties is very limited and is likely to remain so for some considerable time.

2. The Fulcher Equation

In the early 1920s it was found independently by Vogel, Fulcher, and by Tammann and Hesse, that for many glass melts the following equation gives a very good representation of the variation of viscosity with temperature over a wide temperature range

$$\eta = A \exp[B/(T-T_O)] \ , \tag{23}$$

where A, B and T_O are constants for a particular melt. Although German authors honour all its discoverers by referring to it as the VFT equation, elsewhere Fulcher usually receives all the credit. Figure 23 shows how well the equation fits the results for sheet glass over the temperature range from 550° to 1400°C. For many years it was regarded as an empirical equation, but recently it has been recognized as having some theoretical justification. It can be very useful practically for interpolation purposes. If one has good reason to believe that the equation is obeyed for a particular glass, then in principle it should be necessary to measure the viscosity at only three temperatures to obtain the complete viscosity-temperature relationship. Three values are sufficient to determine the three constants A, B and T_O. A more common application is to use it to bridge the gap which is commonly encountered between the viscosity range in which viscosities can readily be measured by the rotating cylinder method and that in which the fibre extension method gives satisfactory results.

F. Flow Properties in the Transformation Range

1. Variation of Viscosity with Time

As explained in Chapter I, the transformation range is that temperature region in which there is a change from solid-like to liquid-like behaviour as the temperature of the glass is raised. It is a region in which changes in glass properties with time occur following a sudden change in temperature. The glass structure is capable of changing in response to the temperature change since the constituent atoms are sufficiently mobile, yet the change in structure is not so rapid as to make it difficult to observe the rate of change in those properties which are struc-

54

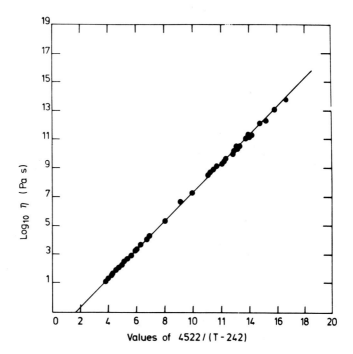

-Fig. 23. Fulcher plot of viscosity results for sheet glass
(Physical Properties Committee, 1956).

ture-sensitive. In this range one can also observe marked differences in the mechanical behaviour of the glass depending on
the rate at which it is subjected to changes in the forces applied to it.

One of the first studies of the change of glass properties
with time in the transformation range was carried out by Lillie
(1933). Some of his viscosity results are shown in Fig. 24.
Both series of measurements were made by the extension of a
fibre method at the same temperature (487°C). A fibre which had
previously been heated for a long time at 478° has a viscosity
which decreases with time to a final constant value. On the
other hand, a fibre which is taken in the as-drawn condition
has a viscosity which increases with time to the same constant
value. Clearly the constant viscosity is the equilibrium viscosity at the temperature of measurement. What we are seeing
here is the effect of slow changes in glass structure to a
structure which is characteristic of the measurement temperature

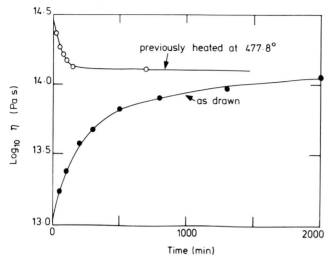

Fig. 24. Change with time of viscosity measured at 487°C.

of 487°C. It is important to remember that much greater changes
in viscosity with time may occur if the glass phase-separates
during the measurement (Simmons et al., 1974). It may not always be easy to determine which part of the observed change is
due to stabilisation and which to phase separation.

2. Stress Relaxation Phenomena

      Some experiments on the resilience of glass by Kirby (1956)
illustrate very clearly and simply the marked effect of the
rate of application of stress on the flow properties of glass.
His apparatus is shown in Fig. 25. A small steel ball bearing is
dropped from a measured height onto the glass which is contained
in a crucible. When the glass is at room temperature, the ball
rebounds almost to the height from which it was dropped. Kirby
measured the effect of glass temperature on the height of re-
bound and found that the glass showed some resilience up to temperatures more than 100°C above the transformation range. In
this experiment the rate of application of stress to the glass
is very high. In fact the time for which the ball is in contact with the glass is of the order of only 20 µs and during
this time, at temperatures within a range of 100°C above the
transformation range, there is insufficient time for the stress

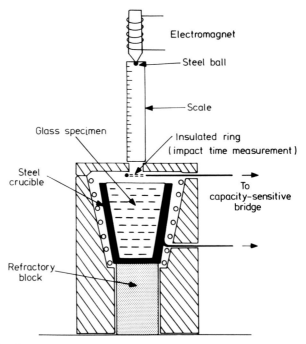

Fig. 25. Kirby's apparatus for measuring the resilience of glass.

produced by the impact of the ball to be reduced significantly by viscous flow. Consequently the glass shows a considerable amount of elastic behaviour.

A mechanical model which behaves in a qualitatively similar way to the glass in this experiment is illustrated in Fig. 26. It consists of a spring, S, connected to a piston, P,

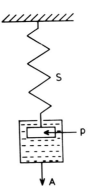

Fig. 26. Maxwell element.

which can move vertically in a cylinder filled with a viscous
oil. The oil-filled cylinder and piston is usually referred to
as a dashpot and the series arrangement of spring and dashpot is
termed a Maxwell model or Maxwell element. If one were to suddenly pull down on A, the spring would stretch instantaneously
and this would then be followed by a slow movement of the piston
in the cylinder. If A were released almost immediately after
being pulled down, the spring would return to its original length
and the movement of the piston in the cylinder would be negligi-
bly small. The assembly would then behave almost as if the
dashpot were absent. If on the other hand a constant force were
applied to A, the piston would eventually move at a constant
speed in the cylinder and the velocity of A would be proportional to the applied force. This simulates the behaviour of
the glass above the transformation range. Under a constant applied stress it behaves as a viscous liquid, but when subjected
to a short impulse it responds as an elastic solid. Mechanical
models of this kind are useful in thinking about the flow properties of materials and in devising equations which describe the
behaviour of materials under various loading conditions.

The Maxwell model may be used to derive an equation for
the rate of release of stress which occurs when glass is held at
a constant temperature in the transformation range. This equation is of practical value in considering the annealing of glass,
a process to which all glass articles are subjected after they
have been formed from the molten material. The article is
brought to a uniform temperature in, or slightly above, the
transformation range. It is held at this temperature for a time
sufficient for any stresses present to decay to an acceptably
low value. The article is then cooled at such a rate that
stresses, which tend to arise as a result of temperature gradients produced by cooling, are not so great as to significantly
reduce the strength of the finished article.

Suppose A in the Maxwell model is moved down suddenly by a
distance l_o and is then held fixed. The instantaneous extension
of the spring will be l_o, thus producing a tension F_o in the
spring given by

$$F_o = K\, l_o \ ,$$

where K is the stiffness of the spring. As time elapses, the
tension in the spring will cause the piston to move upwards in
the cylinder. The extension of the spring will consequently
decrease and so will the force which must be applied to A to
keep the displacement, l_o, constant. At time t let l_o be the
extension of the spring and l_p be the extension in the dashpot,
i.e. the movement of the piston P relative to the cylinder.
Then

$$l_s + l_p = l_o \qquad (24)$$

The velocity of P will be proportional to the force pulling it
upwards and this force is the tension in the spring, which we
shall call F_t. Thus

$$dl_p/dt = F_t/C\eta \ ,$$

where η is the viscosity of the oil in the dashpot and C is a
constant related to the design of the dashpot. On differentiating equation (24) we obtain

$$dl_s/dt + dl_p/dt = dl_o/dt = 0 \ .$$

Hence

$$K^{-1}.dF_t/dt + F_t/C\eta = 0 \ . \qquad (25)$$

On integrating this equation, we obtain

$$F_t = F_o \exp[-(Kt/C\eta)] \qquad (26)$$

The analogue of this equation, in which material properties are
substituted for the spring stiffness and the viscosity of the
dashpot oil, is

$$S_t = S_o \exp[-(Et/\eta)] \qquad (27)$$

where E is the Young's modulus and η is the viscosity of the
material. S_t and S_o are respectively the stresses at time t
and zero respectively, when the material is subjected to a constant strain. Equation (27) is known as the Maxwell equation
of stress release. The ratio η/E has the dimension of time.
It is the time τ in seconds for the stress to decay to $1/e$ of
its value at time zero. τ is called the relaxation time. The
value of τ decreases rapidly with increasing temperatures largely because of the decrease in viscosity. The Young's modulus E
also decreases with temperature but far less rapidly. Assuming
E to remain constant at its room temperature value of approx-
10^{11} Nm^{-2}, the value of τ at a viscosity of 10^{12} Pa s is 10 s.

In Kirby's experiment the contact time between the ball and the glass is approximately 20 μs. A relaxation time of this order corresponds to a viscosity of approximately 10^6 Pa s. Hence some elastic response should be expected at viscosities greater than this value. This corresponds to what Kirby observed.

The simple Maxwell equation is of value for order of magnitude calculations. However it does not accurately describe the relaxation of stress in real glasses. The law of stress release is more complicated and can be modelled by an assembly of Maxwell elements coupled in parallel. There is a wide distribution of relaxation times. The non-exponential nature of the phenomenon of stress release is obvious from the experimental results of Kurkjian (1963) shown in Fig. 27. It is of considerable interest

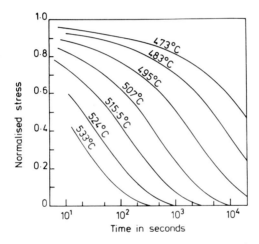

Fig. 27. Effect of temperature on the stress relaxation of a silicate glass.

that these curves, representing results obtained over a range of temperatures, can be superimposed satisfactorily onto a single curve at a chosen reference temperature. This is shown in Fig. 28, the reference temperature chosen being 473°C. The transformation involves a time scaling calculation in which a time t_T corresponding to a given value of S_t/S_o at temperature $T°K$ is converted to a time t_R at the reference temperature $T_R°K$ by use of the equation

$$\log t_T - \log t_R = Q/2.3R(T_T^{-1}-T_R^{-1}) \ . \qquad (28)$$

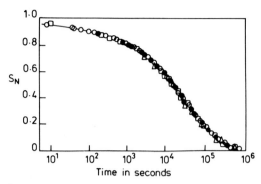

Fig. 28. Data of Fig. 27 replotted for a reference temperature of 473°C.

Q is an activation energy, which was found to be 150 Kcal mole^{-1}. This is identical to the activation energy for viscous flow for the same glass.

These considerations of rates of stress release in the transformation range are of great practical value when dealing with situations in which the glass is being subjected to strains which vary rapidly with time, e.g. when the glass is being thermally toughened (see Chapter IV) and also in some fabrication processes. Using results such as those obtained by Kurkjian, excellent progress has been made in the detailed understanding of the process of thermal toughening. Computer predictions of the variation of stress with time during the process agree very well with those determined experimentally.

G.   Effects of Glass Composition

The viscosity of inorganic glasses is very sensitive to changes in composition. Glasses exist with transformation temperatures ranging from well below 0°C up to approximately 1100°C. Most commercial glass compositions have transformation temperatures in the range 400 to 1100°C.

The literature describing the effects of glass composition on viscosity is extensive. Most of the work on silicate glass compositions up to 1950 has been summarised by Morey (1954). The book by Mazurin et al. (1975) is a very valuable collection of data on two- and three-component oxide glasses and a bibli-

ography published by the International Commission on Glass (1970)
is also a useful source of information.

It is sufficient to give here only a brief indication of
some of the trends observed in silicate and some borate systems.
Figure 22 compares the viscosities of the three glass-forming
oxides, SiO_2, GeO_2 and B_2O_3. The viscosities at any given temperature increase in the order of the melting points (450°, 1116°
and 1723°C respectively and, at the melting points, the values
of the viscosities of the three melts are about the same. As
one might expect from this figure, the glasses based on silica
require much higher temperatures for their manufacture than those
based on B_2O_3. Many borate glasses can easily be made at temperatures below 1000°C. These compositions form very fluid melts
which rapidly become homogeneous. The commercially important
silicate glasses, on the other hand, require melting temperatures of at least 1400°C. Even at these temperatures their viscosities are relatively high and melting times of several hours
are required to make them sufficiently homogeneous.

The addition of small percentages of an alkali metal oxide
to either SiO_2 or GeO_2 greatly reduces the viscosity (Fig. 29)
but in the range of alkali contents encountered in commercial
compositions, the effect of changes in alkali content is far less
great. Figure 30 shows the results of Shartsis et al (1952) for
the variation of viscosity with composition in three binary silicate systems and Fig. 31 shows the same results, together with
others for higher temperatures, plotted in a different way. The
three alkali oxides are now seen to have the same effect. In
Chapter I the structural effect of alkali additions to silica
was described. Each alkali molecule added produces two non-
bridging oxygens, i.e a break in the silica network. It would
appear from the viscosity results that the main factor determining the viscosity of an alkali silicate melt at a given temperature is the number of such breaks per unit volume. This is
a good illustration of the interesting patterns of behaviour
which can sometimes be revealed by plotting property-composition
data in ways which may be suggested by the expected structural
effect of a composition change and by simple ideas as to the way
in which a property may depend on structure. One must guard
against regarding this type of interpretation of property-compo-

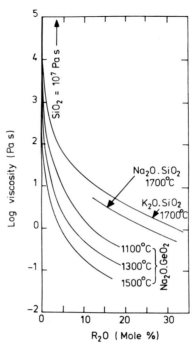

Fig. 29. Viscosities of melts in binary silicate and germanate systems.

sition relationships as an entirely reliable method for investigating the structures of glasses and melts. However, in view of the great difficulty of obtaining structural information about glasses in more direct ways, it is hardly surprising that many glass scientists have devoted considerable time to the analysis of property-composition relationships for the light these relationships may throw on structural changes. It would certainly be unreasonable to take the extreme view that such work is completely misguided and to disregard entirely the evidence obtained from such studies. The book by Scholze (1977) includes a number of discussions of property-composition relationships in simple two-component glasses which illustrate very well this approach to the study of glass structure.

Before leaving this topic, it is interesting to compare the results for the alkali silicate glasses with those in Fig. 32 for alkali borate glasses. Here the results for the Li_2O, Na_2O and K_2O-containing systems are almost identical when the compositions are expressed on a molecular basis. It is known from NMR studies that up to about 30 mol.% of R_2O, the alkali

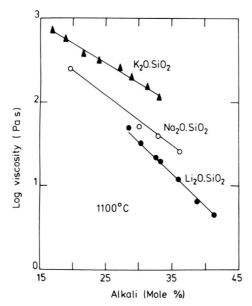

Fig. 30. Effect of composition on the viscosity of alkali silicate melts at 1100°C.

addition results in a change in coordination number of the boron atoms from three to four and no, or very few, non-bridging oxygens are produced, (Bray and Silver, 1960; Bray and O'Keefe, 1963). The variation of viscosity with composition is completely different from that found in the corresponding silicate systems, which is not surprising in view of the difference in the structural effect of adding alkali oxides. The maximum in the curves at about 25 mol.% R_2O is particularly interesting. The fact that it becomes much more pronounced as the temperature falls may be due to a structural change in these melts with changing temperature. Other more direct evidence of such changes is needed to substantiate this point.

The two-component alkali silicate glasses are of little practical interest because of their poor chemical durability and the technologically important silicate glasses contain other constituents which increase the durability and also affect the viscosity. The $Na_2O-CaO-SiO_2$ system is the most important one technologically and many viscosity measurements have been made on glasses in this system. The results are given in Morey's

64

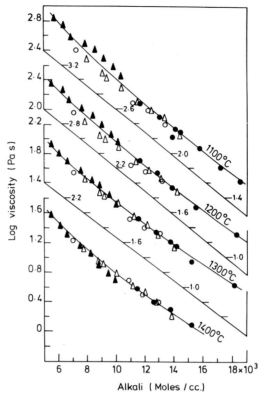

Fig. 31. Effect of alkali concentration in mole/cc on the viscosity of alkali silicate melts. $\triangle$ K_2O, O and $\triangle$ Na_2O, $\bullet$ Li_2O.

book (1954). The substitution of Na_2O by CaO on a molecular basis results in a marked increase in viscosities at all temperatures. Because Na_2O and CaO have similar molecular weights, similar viscosity changes are observed when the substitution is made on a weight percentage basis. Container glass and flat glass compositions also contain significant amounts of MgO and Al_2O_3. It is obviously impossible to discuss, in a simple way, viscosity-temperature relationships in a five-component system. It is also difficult to plan experiments on the effects of composition in such a system.

There are two main reasons for being interested in viscosity-composition relationships in complex multi-component systems. One may wish to optimise a glass composition in the

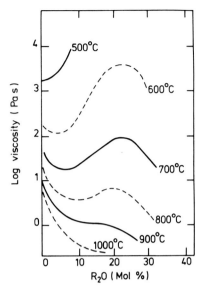

Fig. 32. Effect of composition on the viscosity of alkali
borate melts.

sense of producing a glass with "improved" working properties
for a particular manufacturing process. Thus it may be considered desirable to use a glass with a slightly steeper viscosity-temperature curve in the hope of increasing the production speed. A second and more common need for such information
is that one is always concerned to maintain a supply of glass
of constant viscosity to the forming machines. It is therefore
of prime importance to know how sensitive the viscosity is to
any change in glass composition.

Results of studies in which one component is substituted
for another may not be of much value since in practice one may
be interested in the effect of almost any change in composition
and it is more than likely that the particular composition
change in which one is especially interested may not have been
studied. For dealing with this difficult situation a recent
investigation by Lakatos et al. (1972) is extremely valuable.
They carried out viscosity measurements on 25 glasses in the
system SiO_2-Al_2O_3-Na_2O-K_2O-CaO-MgO. The compositions of the
glasses were chosen to cover the range of commercial container
and sheet glass compositions and also to facilitate subsequent
statistical analysis. The method of multiple regression analysis

was used to obtain equations relating the three constants in
the Fulcher equation* to the glass composition. These equations
are

$$B = -6039.7 \ Na_2O - 1439.6 \ K_2O - 3919.3 \ CaO$$
$$+ \ 6285.3 \ MgO + 2253.4 \ Al_2O_3 + 5736.4$$
$$A = -1.4788 \ Na_2O + 0.8350 \ K_2O + 1.6030 \ CaO$$
$$+ \ 5.4936 \ MgO - 1.5183 \ Al_2O_3 + 1.4550$$
$$T_O = -25.07 \ Na_2O - 321.0 \ K_2O + 544.3 \ CaO$$
$$- \ 384.0 \ MgO + 294.4 \ Al_2O_3 + 198.1$$

The symbols Na_2O, K_2O, etc. represent molecular contents
in relation to 1.0 mole of SiO_2 in the glass composition. The
following table shows how well the viscosity values calculated
using these formulae and the Fulcher equation agree with the
measured values for two standard glasses. The two glasses differ considerably in composition and their viscosities were measured on a number of different viscometers. The agreement with
the calculated values is extraordinarily good.

TABLE V

Comparison of viscosity measurements (η in Pa s) on two standard glasses with values calculated from their compositions

Temperature	Glass A log η		Glass B log η	
	Measured	Calculated	Measured	Calculated
600	10.98	11.00	11.05	11.16
700	8.23	8.28	8.13	8.19
800	6.46	6.51	6.30	6.34
900	5.22	5.26	5.05	5.08
1000	4.32	4.35	4.14	4.16
1100	3.62	3.64	3.45	3.46
1200	3.07	3.08	2.91	2.92
1300	2.63	2.62	2.47	2.48
1400	2.20	2.24	2.11	2.12

Glass A is a standard sheet glass [Physical Properties Committee S.G.T.
(1956)] and glass B is N.B.S. glass 710 (Napolitano and Hawkins, 1964).

* Expressed in the form $\log_{10} \eta = -A + B/(T - T_0)$.

CHAPTER III

THERMAL EXPANSION

A. The Shape of the Thermal Expansion Curve

Glasses, like most other solids, expand on heating (except
at very low temperatures when some contract). Whilst the shape
of the thermal expansion curve is almost linear for the great
majority of crystalline solids, the expansion curve of a glass
shows a more or less pronounced increase in slope at the transformation temperature. This is shown in Fig. 1 and the reason
for the change in slope is discussed in Chapter I.

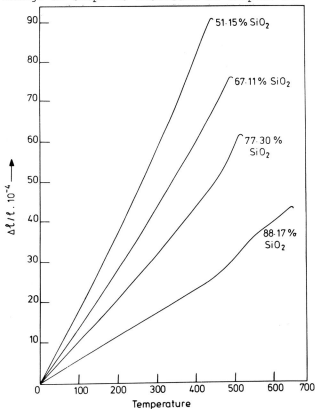

Fig. 33. Thermal expansion curves of Na_2O–SiO_2 glasses
(Turner and Winks, 1930b).

The magnitude of the change in slope in the transformation
range varies considerably from one glass to another. This is

illustrated in Fig. 33 which shows the thermal expansion curves for a series of glasses in the Na_2O-SiO_2 system.

The apparent contraction at the high temperature end of the expansion curves is a result of the way in which the measurement is made. As will be described later, a thermal expansion measurement is usually carried out on a rod of glass which is under axial compression whilst the measurement is being made. As a result of the decrease in viscosity with increasing temperature, a temperature is eventually reached when the glass begins to flow under the applied compressive stress. The measurement is terminated at this point. The viscosity at which flow becomes detectable in the specimen depends upon the magnitude of the compressive stress applied and this varies somewhat from one apparatus to another. Typically, the viscosity is about $10^{11.5}$ Pa s. The temperature corresponding to this viscosity is termed the dilatometric softening point and is denoted by the symbol Mg.

The shape of the expansion curve of a glass is affected, to an extent dependent on the glass composition, by the heat treatment which it receives before the measurement is carried out. Borosilicate glasses are particularly sensitive to heat treatment effects as Fig. 34 shows.

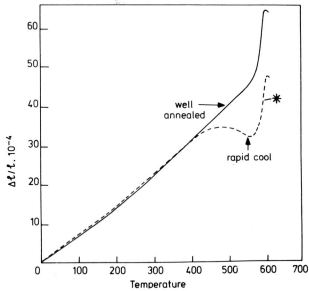

Fig. 34. Effect of quenching on the thermal expansion of a borosilicate glass (Turner and Winks, 1930a).

The rapidly quenched specimen begins to contract as the transformation range is approached. The glass has a relatively high fictive temperature and therefore a lower density than if it had been slowly cooled. The contraction is due to the structure changing towards a more dense form characteristic of the equilibrium configuration. This densification effect may be so marked as to outweigh the normal expansion due to the increasing amplitude of atomic vibrations, which occurs as the temperature rises. When the glass has attained the equilibrium configuration, it then continues to expand in a normal manner. These observations make it clear that when measuring the thermal expansion curve of a glass it is necessary to subject the specimen to a heat treatment prior to the measurement, in which it is slowly cooled from above the transformation temperature. When determining the expansion curve of a particular commercial glass for quality control purposes it is essential that all samples should be given a standardised heat treatment before the measurement is made.

B. Methods of Measurement

Many types of apparatus for measuring the thermal expansion of glass have been described in the literature. Automatic self-recording equipment can be obtained commercially. However, much of the most careful early work on glasses was done with very simple equipment. The most commonly used type of apparatus is the fused-silica dilatometer. In effect, one measures the expansion of a rod specimen of glass relative to that of an equal length of silica glass which is being heated at the same rate. Silica glass has a thermal expansion coefficient which is much less than that of the majority of commercial glasses (5×10^{-7} °C^{-1} compared with a value of $80\text{-}90 \times 10^{-7}$ °C^{-1} for container glass and flat glass compositions). To obtain the absolute expansion of the specimen, a correction must be applied to the measured value for the expansion of the silica glass over the same temperature range. The expansion of silica glass is known with sufficient accuracy (Oldfield, 1964).

Figure 35 shows a simple form of silica dilatometer of the type used in some of the earliest accurate measurements of the thermal expansion of glasses (Turner and Winks, 1928). The

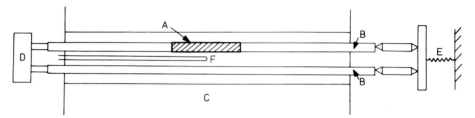

Fig. 35. Silica dilatometer.

furnace is heated at a constant rate, usually between 2 and 3°C
per minute.   It is important that the specimen and the comparison rod of silica should be at the same temperature throughout
the measurement and that the thermocouple recording the temperature should give an accurate record.   This can be achieved by
placing the thermocouple junction so that it is in good thermal
contact with the specimen or by designing the apparatus in such
a way that the thermocouple is in a similar thermal environment
to that of the specimen.   Temperature gradients along the length
of the specimen may be reduced by lining the central section of
the furnace with a thick-walled tube of copper or silver.

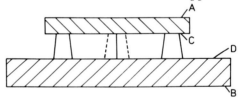

Fig. 36. Interferometer method for measuring thermal
expansion.

An interferometric technique developed at the National Bureau of Standards (Work, 1951) has the advantage that it is an
absolute method and no corrections for silica expansion have to
be applied.   The optical flats A and B (Fig. 36) are separated
by three carefully ground pyramids made from the glass under investigation.   As the glass expands, the spacing between the faces
C and D increases.   This is observed by the movement of interference fringes across the field of view of a telescope.   Interferometer methods using a laser as a light source have also been
used for measuring the expansion properties of glasses and glass-
ceramics having expansion coefficients which are as low as, or
even lower than that of silica glass (Plummer and Hagy, 1968).

71

C. Effects of Glass Composition

The thermal expansion coefficient of oxide glasses ranges
from almost zero for glasses in the TiO_2-SiO_2 system to over
200×10^{-7} $°C^{-1}$. The expansion coefficients of common metals
cover a rather similar range of values, as Table VI indicates.

TABLE VI

Thermal Expansion Coefficients of some Metals near Room
Temperature

	10^{-7} $°C^{-1}$
Aluminium	230
Copper	167
Nickel	128
Iron	120
Platinum	89
Tungsten	45

Approximate values of expansion coefficients of three important
one-component oxide glasses are given in Table VII.

TABLE VII

Expansion Coefficients of One-Component Oxide Glasses

	10^{-7} $°C^{-1}$
SiO_2	5-6
GeO_2	77
B_2O_3	150

The value of the expansion coefficient of a glass depends
on the temperature range over which it is calculated. Below Tg
the gradient of the expansion curve increases slightly with increasing temperature. Below 0°C the expansion coefficient of
some oxide glasses varies markedly with temperature as is shown
by the results of Krause and Kurkjian (1968) in Fig. 37. Vitreous silica shows a negative expansion coefficient over a wide
temperature range up to about 150°K. All the glasses in the
figure which show a negative thermal expansion coefficient at
very low temperatures are either known to have, or are believed
to have, a silica-like structure, i.e. the structure is built
from tetrahedra joined at their corners.

The addition of alkali metal oxides to silica to give the

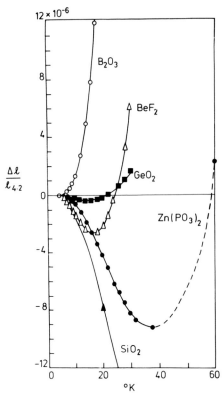

Fig. 37. Thermal expansion curves of some glasses at low temperatures.

alkali oxide-silica series of glasses results in a marked increase in thermal expansion which increases almost linearly with
the alkali content (Fig. 38). Alkaline earth oxides such as MgO
and CaO also increase the expansion coefficient, but not so
markedly. Thus replacement of an alkali oxide by an equivalent percentage of an alkali earth oxide reduces the expansion
coefficient. The alkali borate glasses show a quite different
pattern of behaviour with a minimum expansion coefficient at
about 20 mol% R_2O (Fig. 39).

Similar maxima or minima are observed in graphs showing the
variation of other physical properties of alkali borate glasses
with composition. This behaviour, when contrasted with the
rather simple pattern found in the silicate systems, led to the
introduction of the term "the boric oxide anomaly" and to various attempts to explain it in terms of the change of the co-

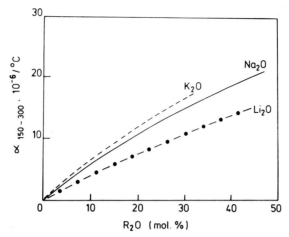

Fig. 38. Thermal expansion coefficients of glasses in the system R_2O-SiO_2.

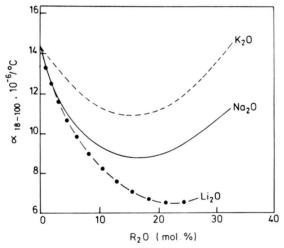

Fig. 39. Thermal expansion coefficients of glasses in the system R_2O-B_2O_3.

ordination number of the boron atoms from 3 to 4 as the R_2O content is increased. However, if there is any relationship between expansion coefficient and co-ordination number, it cannot be a simple one since it is now known that the change of co-ordination number with the addition of R_2O is continuous over the range of compositions up to about 30 mol% R_2O (see Chapter II, section G). Also, whilst it is believed that changes in co-ordination number occur in the Na_2O-GeO_2 system and there is a pronounced maximum in the refractive index of these glasses at about 15 mol% Na_2O, the expansion coefficient of the glasses in this system increases almost linearly with increasing Na_2O content (Mazurin et al., Vol. II, p. 367, 1975).

Lakatos, Johansson and Simmingsköld (1973) have derived an equation for calculating the expansion coefficient from the composition. This is deemed to be satisfactory within the limited range of container and sheet glass compositions. It was derived from a statistical analysis of the results obtained on a carefully planned series of experimental glasses. Their work is a considerable improvement on previous attempts to derive expansion formulae, both with respect to the planning of the experiments and to the analysis of the results.

The formula gives the mean expansion coefficient over the temperature range from 20° to 300°C. It is:

$$\alpha = 5.13 + 21.0864\ Na_2O + 27.5584\ K_2O +$$
$$1.38887\ CaO - 2.3930\ MgO - 8.8638\ Al_2O_3 \qquad (30)$$

The expansion coefficient is given in units of $10^{-6}\ °C^{-1}$. The oxide formulae in the equation represent the number of moles of the oxide per mole of SiO_2 in the glass composition.

The agreement with experimental values is very good. Thus the mean value of the expansion coefficient of a standard sheet glass determined by a number of laboratories was $81.5 \pm 1.5 \times 10^{-7}\ °C^{-1}$ relative to vitreous silica (Physical Properties Committee, 1956). This corresponds to an absolute value of $87 \times 10^{-7}\ °C^{-1}$. The value calculated from the composition using Equation 30 is $87.75 \times 10^{-7}\ °C^{-1}$, an agreement well within the quoted error.

D. Stresses in Glass Seals

Whenever a glass is fusion-sealed to another material the thermal expansion properties of the two materials must be carefully chosen to ensure that the glass does not crack when the seal is cooled. Seals between glasses of differing expansion coefficient and between glasses and metals are widely used in the manufacture of electric lamps and electronic valves. Although small radio receiver valves have now been completely replaced by their semi-conductor equivalents, a large number of special valves incorporating glass are still manufactured for

various engineering applications, and also for one very well
known domestic application. One of the largest and most complex glass-envelope devices is the colour TV tube.

It is difficult to imagine how the modern electronics industry could have come into being, or indeed how modern atomic
physics could have developed, if glass had not been available.
The combination of the transparency of glass, its relatively
low permeability to gases, and the ease with which it can be
fused to metal to make vacuum tight seals, were of vital importance in the relatively simple experiments, carried out late in
the nineteenth and early in the twentieth century, which led to
the discovery of the electron and the measurement of its properties.

The technology of the design and manufacture of glass to
metal seals is well described in a number of texts (Partridge,
1949; Kohl, 1972; Henderson and Marsden, 1972; Espe, 1968).
In this section only the most elementary account will be given
of the factors which determine the stresses in such a seal.

The control of stresses by suitable choice of expansion
properties is also important in another major field of application, i.e. vitreous enamelling. Vitreous enamels contain a
high percentage of glass phase, but in addition most enamels
also contain crystalline materials introduced to produce colour
or the required degree of opacity. The technology of these materials and their application has been described by Andrews
(1935) and Parmalee (1951).

A perfect thermal expansion match between the materials
which are the components of a glass seal is never possible.
One has always to contend therefore with some stress in the seal.
Although it is possible to choose combinations of glass and metal in which the expansion curves of the two materials are closely
similar over the entire temperature range from room temperature
up to the annealing temperature of the glass, batch to batch
variations in the compositions of the materials occur which
affect the expansion properties. It is therefore impossible to
guarantee that any seal will be free from stress. Because the
thermal expansion properties of glasses used in sealing applications have such a marked effect on the seal stresses, it is
essential to maintain a strict system of quality control to en-

sure that the expansion properties are always within specification. Theories have been developed which allow the stresses in a number of important simple designs of seal to be calculated. These designs are illustrated in Fig. 40.

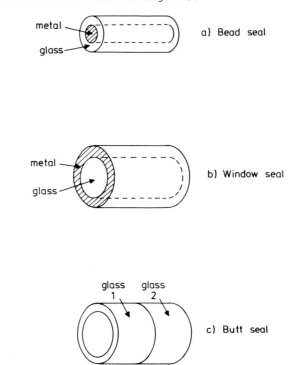

metal

glass

a) Bead seal

metal

glass

b) Window seal

glass glass
  1   2

c) Butt seal

Fig. 40. Simple designs of glass seal.

The stress theories do not attempt to provide a full analysis of the stress system. Usually the more complicated parts of the stress system near the ends of the seal are ignored, although a few publications exist which deal analytically with these stresses. With the advent of the computer and the development of numerical methods of stress analysis there is now no reason why such complex problems should not be solved relatively easily.

To illustrate the general approach to the calculation of seal stresses, the simple example illustrated in Fig. 41 will be analysed. It is usually called a sandwich seal and consists of a sheet of material A sealed between two equally thick sheets of material B. Both sheets are assumed to be of infinite extent in directions parallel to the seal interface. This assumption

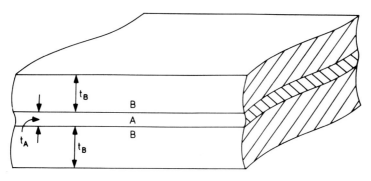

Fig. 41. Sandwich seal.

eliminates the need to consider end effects. Sandwich seals are
used for material control purposes when material A is frequently
a metal alloy and material B is a glass. It will be assumed
initially that both materials have linear expansion curves, i.e.
that the thermal expansion coefficients of the two materials
α_A and α_B, are independent of temperature (Fig. 42). The seal

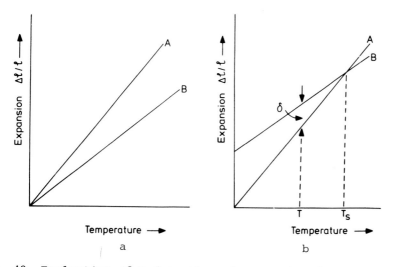

Fig. 42. Evaluation of δ from thermal expansion data.

is made by heating the components in contact to a temperature
well above the annealing temperature of the glass so that the
glass viscosity is low enough for it to flow and wet the metal.
The seal is then cooled. At temperatures well above the glass
annealing temperature no stress is set up on cooling even though

the glass contracts at a rate different from that of the metal.
The glass viscosity is too low for any permanent stress to arise.
However, when the glass viscosity reaches a value of approximately 10^{12} Pa s, the rate of stress release in the glass by viscous flow is such that some stress begins to form. As the temperature falls further the glass becomes effectively an elastic
solid so that no significant further stress release is possible.
It will be assumed for the purpose of developing equations for
the stress distribution that there exists a well-defined temperature T_s above which no stress can be established in the glass
and below which the glass behaves as an ideal elastic solid.
This will be referred to as the setting point. In reality, of
course, the transition is a gradual one. Also the value of the
effective setting point T_s depends to some extent on the rate at
which the seal is cooled. The higher the cooling rate, the higher
the effective value of T_s.

Figure 42b shows the expansion curves of the two materials
redrawn in such a way that they intersect at T_s. It is required
to calculate the stresses at some lower temperature T. If the
two materials had not been sealed together the free contraction
of A per unit length in directions parallel to the interface
would have been $\alpha_A(T_s-T)$ and the corresponding free contraction
of B would have been $\alpha_B(T_s-T)$. There would then have existed a
difference in dimensions parallel to the interfaces of
$(\alpha_A-\alpha_B).(T_s-T)$. This quantity is called the differential free
contraction and this will be denoted by δ. Since the sheets are
sealed together this difference in dimensions cannot exist. At
the sealing interface at least, the two sheets must contract together. Moreover, since the sheets are of infinite extent, con-
tractions in directions parallel to the interfaces must be constant throughout the seal thickness. Expressed in another way,
any plane section through the sheet, which lies in a direction
at right angles to the plane of the sheets, must remain plane
as the sheets cool below T_s. This is necessary if continuity
is to be maintained within the material of each component of
the seal.

If one is dealing with a situation where δ is positive,
i.e. α_A is greater than α_B, the requirement that the sheets contract together results in tensile stresses being set up in sheet

A and compressive stresses in the two sheets of material B. Let σ_A be the stress in sheet A and σ_B the stress in sheet B. The strains e_A and e_B corresponding to the stresses are respectively

$$e_A = \sigma_A(1-\nu_A)/E_A$$

and

$$e_B = \sigma_B(1-\nu_B)/E_B$$

where E is the Young's modulus and ν the Poisson's ratio. To simplify the algebra, we shall assume in the following that ν_A and ν_B are equal.

Thus for each sheet, the change in dimensions on cooling from T_S to T is made up of two components, a component due to the temperature change and one due to the stress. The sum of these two components must be the same within each sheet. Thus

$$\sigma_A(1-\nu)/E_A - \alpha_A(T_S-T) = \sigma_B(1-\nu)/E_B - \alpha_B(T_S-T)$$

or

$$(\sigma_A/E_A - \sigma_B/E_B)(1-\nu) = \delta \ . \tag{31}$$

This equation gives one relationship between the stresses. To calculate either stress, a second equation is needed. This is obtained by considering the balance of forces acting across an imaginary plane cut through the seal in a direction at right angles to the interface (Fig. 43). For equilibrium there must

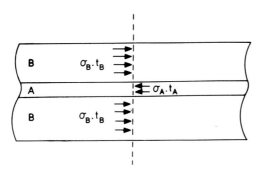

Fig. 43. Balance of forces within a sandwich seal.

be no net force acting across this section, i.e. the internal forces due to the stresses in the seal components must be balanced. If we consider unit thickness of the seal measured in a direction normal to the plane of the paper, the force arising from the stress in sheet A is $\sigma_A.t_A$ whilst in each sheet B of material B there is a force $\sigma_B.t_B$. From the condition of equilibrium

$$\sigma_A t_A + 2\sigma_B t_B = 0 . \qquad (32)$$

Combining equations 31 and 32 we obtain

$$\sigma_B (1/E_B + 2t_B/(t_A E_A))(1-\nu) = -\delta \qquad (33)$$

From this σ_B can be calculated and then σ_A by use of Equation 32.

Thus the stresses are directly proportional to the differential free contraction and they depend only on the ratios of the sheet thicknesses, not on the actual value of these thicknesses.

When t_B is very small compared with t_A, i.e. we have a thin coating of glass on each side of material A, almost the whole of the differential free contraction strain is carried by the coating layers. σ_B then has its maximum value of $-E_B\delta/(1-\nu)$. When dealing with a situation of this kind, it is very desirable that the expansion coefficient of the glass, α_B, should be less than that of the substrate. The stresses in the coating are then compressive. As we shall see in Chapter IV, glass can withstand very high compressive stresses. Indeed, if one could be sure that no tensile stresses could develop in the coating, the value of δ could be very high without the risk of cracking the glass. In vitreous enamelling and when glazing ceramics the aim is always to ensure that the glassy coating has a lower expansion coefficient than the substrate material.

A widely used rule of thumb in the design of glass to metal seals is to ensure, by suitable choice of dimensions and material properties, that no calculated stress is greater than 10 MNm^{-2}. This value is at least five times smaller than the stress at which the glass may fracture. The use of such a generous factor of safety is advisable bearing in mind the difficulty of calculating stresses near the ends of the seal, which are usually higher than those in the central regions.

The geometrical construction used to determine δ which was illustrated in Fig. 42 is not limited by the assumption made earlier that the expansion coefficients are independent of temperature. Figure 44 shows the same construction used in a situation where neither expansion curve is linear. Note that δ passes through zero and changes sign as the seal is cooled. As a result, those stresses which were tensile become compressive and vice versa.

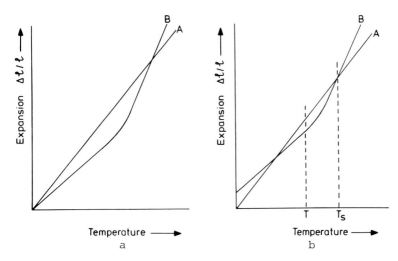

Fig. 44. Evaluation of δ from thermal expansion data.

Using this construction and the appropriate stress theory
it is possible to calculate how the stresses in a seal will vary
with temperature. It is important to have this information
since it is quite possible to have situations in which the
stresses in a seal are acceptably low at room temperature but
are excessive at some temperature between room temperature and
T_s. This use of the thermal expansion curves to assess a sealing combination is not a very satisfactory procedure, partly
due to the uncertainty in fixing the value of T_s. The method
which is most generally used in practice for assessing a combination of sealing materials is to make a simple seal design
of standardised dimensions such as a sandwich seal and to measure by photoelastic methods the way in which the stress in the
seal varies with temperature as it is cooled from above the
annealing temperature. Comparison of the stress-temperature
curve obtained in this way with those obtained with the same
standard seal design but using combinations of materials which
are known to be satisfactory provides a reliable guide as to
the suitability of any new combination. The same technique is
used for quality-control purposes in the manufacture of sealing
glasses and sealing alloys.

It is interesting that alloys can be made which have an
expansion curve very similar in shape to that characteristic of

a glass, i.e. they show an increase in slope above a certain
transition temperature. By changing the alloy composition one
can modify the slopes below and above the transition temperature
and also change the transition temperature. The increase in expansion coefficient at a particular temperature is associated
with a loss in ferromagnetism in the alloy, i.e. it occurs at
the Curie temperature. Sealing alloys which show this type of
behaviour are based on the nickel-iron system. They are readily
available in the form of sheet and tubing and are widely used
in a great variety of seal designs.

Although the maximum difference in expansion coefficients
is limited to about 5×10^{-7} $°C^{-1}$ in the majority of seal designs,
ingenious designs have been developed which permit materials
having a much greater difference in expansion coefficients to be
sealed together. Thus molybdenum with an expansion coefficient
of 55×10^{-7} $°C^{-1}$ can be sealed to vitreous silica, and copper
with an expansion coefficient of 180×10^{-7} $°C^{-1}$ can be sealed
to a borosilicate glass with an expansion coefficient of 32×10^{-7} $°C^{-1}$. In the molybdenum-silica seal, which is widely used
in a large range of lamps made with silica glass envelopes, the
current-carrying lead is a very thin rectangular strip of metal,
the edges of which have been feather-edged by an electrolytic
treatment. For high power lamps used in film and TV studio
lighting a large number of these seals are used in parallel to
provide the necessary current carrying capacity. In the compact
source lamps used in car headlamps and home film projectors,
quite a simple seal design is satisfactory. The same principle
is used in the copper seals referred to. These usually take the
form of a copper tube, one end of which is sealed into the glass.
This end is machined to a feather edge to reduce the rigidity of
the metal. The fact that copper is ductile also helps to prevent
excessively high stresses being set up in the glass.

E. Thermal Stresses

Glass articles are frequently subjected to sudden heating
or cooling. This produces temperature gradients in the material.
Consequently stresses are produced as a result of different parts
of the glass expanding or contracting by different amounts. It
is important to be able to calculate these stresses, or at least

have an understanding of the factors which determine their magnitude. The thermal stresses may be so high as to fracture the
glass, this being particularly likely to happen if the temperature gradients produce tensile stresses in the glass surface.

The methods used to calculate thermal stresses are very similar to those used for calculating the stresses in glass seals.
A very simple example will be considered first. Suppose we have
a sheet of glass at a uniform temperature T_i and that both surfaces of the sheet are suddenly cooled to a lower temperature T_o.
The surface layers, if they were free to do so, would contract
by an amount $\alpha.(T_i-T_o)$ per unit length. However they are pre-
vented from contracting by the rest of the sheet which is still
at the initial temperature. The surface layers are therefore
subjected to a tensile strain in directions parallel to the plane
of the sheet. The surface tensile stress σ_o is given by the
equation

$$\sigma_o = E\alpha(T_i-T_o)/(1-\nu) \ . \tag{34}$$

For a sheet glass composition approximate values of the physical
properties are

$$E = 0.6 \times 10^{11} \ Nm^{-2}$$
$$\alpha = 90 \times 10^{-7} \ {}^{\circ}C^{-1}$$
$$\nu = 0.23$$

Thus a surface temperature shock of 100°C would produce a surface
tensile stress of 70 MNm^{-2}. This could be sufficient to cause
fracture. A sudden increase in surface temperature would not be
dangerous provided it was uniform over the whole surface of the
sheet since the surface stresses produced would be compressive.
Instantaneous surface temperature changes do not occur in practice, even when a hot glass article is suddenly plunged into
cold water. A thermal barrier layer is produced in the fluid
in contact with the hot surface and this sets a limit to the
rate at which heat can be extracted from the glass. However,
the simple calculation given indicates the order of magnitude
of the stresses which are to be expected.

If the temperature distribution in an elastic solid is known
then, in principle, it is possible to calculate the stress distribution which is produced. This will be illustrated by considering the simplest possible example, a sheet of glass cooling

at a uniform rate of R °C s⁻¹ from an initial uniform temperature.
It will be assumed that the sheet is cooling symmetrically, i.e.
heat is being extracted at the same rate from each of the two
surfaces. Thus at any instant the temperature distributions will
be symmetrical about the central plane of the sheet. The temperature distribution must satisfy the Fourier equation for non-
steady heat conduction in one dimension

$$\partial T / \partial t = a. \partial^2 T / \partial x^2 .$$

The distance x is in the direction normal to the plane of the
sheet, i.e. it is the direction of heat flow. It is convenient
to take the central plane of the sheet as the plane x = 0. The
constant a is the thermal diffusivity of the glass and is given
by the equation

$$a = \lambda / c_p . \rho ,$$

where λ is the thermal conductivity, c_p is the specific heat and
ρ is the density.
The solution of the Fourier equation which applies when every
point in the sheet is cooling at the same rate is

$$T_x = T_0 - Rx^2/2a . \qquad (35)$$

This is a parabolic distribution as shown in Fig. 45. T_0 is the
temperature at the central plane and T_x is the temperature at
distances x from the centre.

The mean temperature of the sheet T_m at any instant can be
calculated from Equation 35. Its value is given by

$$T_m = T_0 - RL^2/24a , \qquad (36)$$

where L is the sheet thickness.

Suppose the sheet were at the uniform temperature T_m and
the temperature distribution given by Equation 36 were suddenly
produced within the sheet, the mean temperature remaining at T_m.
If one were to imagine the sheet divided into very thin layers
in a direction parallel to the surface, each layer would expand
or contract by an amount $\alpha . (T_x - T_m)$. Since the layers are not
free but are bonded together, all the layers must expand or contract by the same amount. Hence each layer will be under stress
in directions parallel to the surface. Let σ_x be the stress at
the distance x from the central plane. This stress will produce
a strain $\sigma_x (1-\nu)/E$. The condition of continuity requires that
the total strain should be constant throughout the sheet, the

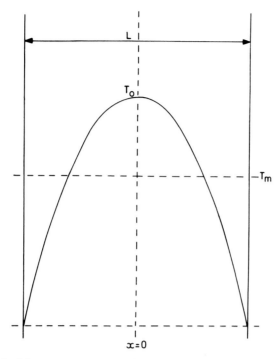

Fig. 45. Parabolic temperature distribution due to uniform cooling of a sheet.

total strain being the sum of the strain due to the temperature change and that due to the stress. Call this total strain e_t. Then

$$e_t = \sigma_x(1-\nu)/E + \alpha(T_x-T_m) \; . \tag{37}$$

In order to obtain the value of e_t, we first integrate both sides of this equation over the thickness of the sheet. This gives

$$e_t \, L = (1-\nu).E^{-1}.\int_{-L/2}^{+L/2} \sigma_x dx + \alpha \int_{-L/2}^{+L/2} (T_x-T_m)dx \; . \tag{38}$$

Now consider the two integrals on the right hand side of this equation. The first represents the net force due to the internal stresses which act across an imaginary plane cut through the sheet in a direction at right angles to this surface. Since the stress system must be in internal equilibrium, this force is zero. The second integral is also zero. It is equivalent to the sum of the deviation of a series of values of T_x from the mean, T_m. Hence e_t is zero. The stress distribution can now be obtained immediately from Equation 37,

$$\sigma_x = -E\alpha(T_x-T_m)/(1-\nu) \ .$$

Thus the stress at the central plane, $x = 0$, is given by

$$\sigma_O = -E\alpha RL^2/24(1-\nu)a \ . \qquad (39)$$

That at the two surfaces, $x = \pm L/2$ is

$$\sigma_S = +E\alpha RL^2/12(1-\nu)a \ . \qquad (40)$$

The surface stresses are therefore positive or tensile stresses whilst that at the centre is compressive.

F. Annealing

The equations obtained in the previous section may be used to calculate the magnitude of the stresses which remain in the glass after it has been cooled at a rate of $R°C \ s^{-1}$ from a temperature above the annealing temperature i.e. from a viscosity less than 10^{12} Pa s. This enables one to decide whether a particular cooling rate will result in permanent stresses being set up in the glass which are acceptably low for the intended application. Figure 46 shows the general shape of the temperature-time sched-

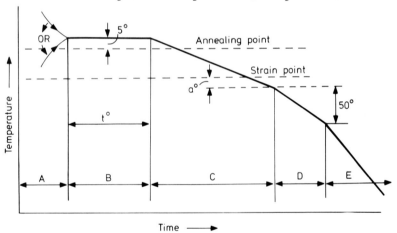

Fig. 46. Temperature-time schedule used for the commercial annealing of glassware.

ule used for annealing glass. The glass is first brought to a uniform temperature throughout, the temperature being such that stresses in the glass, which may already be present, can be released fairly quickly (within minutes). It takes some time for

the entire article to attain a uniform temperature. This depends upon the rate of heat transfer from the surrounding air
in the annealing furnace to the glass surface and also on the
rate of removal by thermal conduction of temperature gradients
in the glass. Knowing the thermal diffusivity of the glass and
the heat transfer coefficient for natural convection at the
glass surface, it is possible, using the Heisler charts to be
found in textbooks on heat transfer (Jakob, Vol. I, p. 284,
1949), to determine the time necessary to bring the glass to a
substantially uniform temperature. A time of the order of 30
min is required for glass 10 mm thick.

   We have already seen in Chapter II how to estimate the time
taken for stress to decay by viscous flow to an acceptably low
level. At a viscosity of 10^{12} Pa s a time of 1000 s should be
sufficient. This time can be reduced by using a higher temperature, but this increases the risk of the glass deforming under
its own weight.

   The glass is now cooled slowly (phase C of the process).
If we are dealing with the simple case of a sheet of glass,
cooling at a rate of $R°C$ s^{-1} will eventually result in the setting up of a parabolic temperature distribution as discussed in
the previous section. However, above the annealing temperature
no stresses will be produced by such a temperature distribution.
Any stress will be relieved by viscous flow. Since the cooling
is started when the glass viscosity is already high, we can expect that the stresses produced by the temperature gradients
will not be relieved completely, in which case small tensile
stresses will be produced in the surface layers balanced by compressive stresses in the central layers. As the glass cools,
its viscosity will increase until eventually the viscosity
throughout the thickness is so high that the material is effectively solid. This can be taken as being when the viscosity is
greater than $10^{13.5}$ Pa s (the strain point), i.e. when the stress
relaxation time is of the order of 10^{4} s. As the sheet continues to cool, no further stresses will be produced so long as
every point within the glass continues to cool at the same rate
$R°C$ s^{-1}. Additional strains, and hence stresses, will arise only
when we depart from this situation and have different rates of
cooling at different points in the sheet. This will result in

differences in the rates at which the various layers in the glass
are trying to contract. Thus the major part of the stress system develops as room temperature is approached and as the glass
is brought to a uniform temperature. At this stage, the rate of
cooling of the surface layers is reduced relative to that at the
centre. If one were dealing with a situation in which no stresses
had been produced in the initial stages of cooling, whilst the
glass was in the annealing range, then the stresses which would
develop at   room temperature would be those resulting from the
removal of the temperature distribution given by Equation 35,
i.e. the stresses would be those given by Equations 39 and 40
but with opposite signs. The surface stresses would be compressive and the central stresses tensile. These equations may
therefore be used to make order of magnitude calculations of the
stresses remaining after annealing at a specified rate of cooling.

One sees from the equations that the stresses depend partly
on the physical properties of the glass and partly on the thickness. For the majority of commercial silicate glasses, the values of the Young's Modulus, E, and Poisson's ratio do not differ
very much from one glass to another. The property which has the
greatest effect on the permissible rate of cooling is the expansion coefficient, which does vary considerably with composition within the range of commercial glasses. The equations
show that the other important factor is the thickness of the
glass. Thus the greater the thickness and the higher the thermal expansion coefficient, the slower must be the rate of cooling.

Figure 46 shows that once the glass is below the strain
point, the rate of cooling is increased considerably. This can
be done without affecting the level of permanent stresses in
the glass after cooling has been completed. Once one is dealing
with an elastic solid, any stresses produced by temperature
gradients disappear when those thermal gradients are removed.
The permanent stresses are determined only by the rate of cooling through the annealing range. When the entire thickness of
the glass is below the strain point, the rate of cooling should
be increased in order to complete the process in the shortest
possible time. The maximum rate of cooling which can be per-
mitted in stages D and E of the process is determined by the re-

quirement that the temporary stresses produced in this stage are
not so high that the glass fractures.

Table VIII, taken from Shand (1958), gives recommended
annealing schedules for the annealing of commercial glassware.
Since glass articles are of irregular shape, the recommended
cooling rates are considerably lower than those which may be
calculated using the equations of the previous section.

G. Measurement of Internal Stresses in Glassware

It is, fortunately, relatively easy to measure the internal
stresses in glass and hence to determine whether an article of
glassware has been adequately annealed or whether the stresses
in a glass seal are acceptable. The method of measurement which
is used almost universally depends upon the effect of the stresses
on the velocity of propagation of polarised light through the
glass, i.e. on the phenomenon known as the photoelastic effect.
For a detailed account the reader is referred to textbooks on
optics (e.g. Tenquitst et al., 1969). There are also a number
of texts which describe the use of the technique in engineering
design (Frocht, 1948).

The velocity of propagation of a plane polarised light wave
in glass which is under stress depends on the direction of polarisation of the light relative to the direction of stress. In
general a plane polarised wave is split into two plane polarised
components as the light enters the glass. These have different
velocities in the glass so that when they emerge there is a
phase difference between them. Interference effects between the
two waves can be produced which result in variations in light
intensity or, if white light is used, give colours. From these
optical effects, the phase difference between the two waves can
be measured. This gives a measure of the magnitude of the stress
in the glass, since the phase difference is proportional to the
stress which produces it.

The simplest example to consider is a block of glass under
uniaxial tension or compression (Fig. 47). The figure shows a
beam of plane polarised light entering the glass in a direction
at right angles to the stress direction and with its plane of
polarisation making an angle of θ with this direction. The light
used is monochromatic. On entering the glass, the wave is re-

TABLE VIII

Commercial annealing schedules. The symbols refer to Fig. 46

Expansion Coeff. of Glass per °C	Thickness of Glass mm	Cooling on One Side						Cooling on Two Sides					
		A Heat Rate °C per Min	B Time t, Min	C Temp. a, °C	C Cool Rate °C per Min	D Cool Rate °C per Min	E Cool Rate °C per Min	A Heat Rate °C per Min	B Time t, Min	C Temp. a, Min	C Cool Rate °C per Min	D Cool Rate °C per Min	E Cool Rate °C per Min
33 x 10⁻⁷	3.2	130	5	5	12	24	120	400	5	5	39	78	400
	6.3	30	15	10	3	6	30	130	15	10	12	24	130
	12.7	8	30	20	0.8	1.6	8	20	30	20	4	6	30
50 x 10⁻⁷	3.2	85	5	5	8	16	85	260	5	5	26	52	260
	6.2	21	15	10	2	4	21	85	15	10	8	16	85
	12.7	5	30	20	0.5	1.0	5	21	30	20	2	4	21
90 x 10⁻⁷	3.2	50	5	5	4	8	50	140	5	5	14	28	140
	6.3	11	15	10	1	2	11	50	15	10	4	8	50
	12.7	3	30	20	0.3	0.6	3	11	30	20	1	2	11

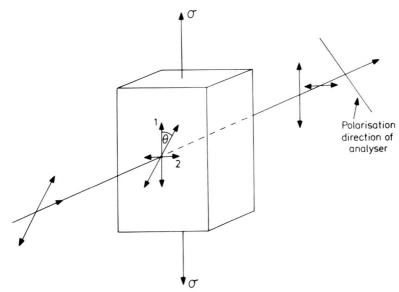

Fig. 47. The photoelastic effect.

solved into a component 1 with its polarisation direction in
the direction of stress and a second, 2, with its polarisation
direction at right angles to this. If A is the amplitude of
the incident wave, the two components have amplitudes A cos θ
and A sin θ respectively.

In the following it will be assumed that θ = 45°, when the
two components have equal amplitude. **Which of** the two components has the higher velocity depends on whether the stress is
tensile or compressive. If it is tensile, the wave polarised
in the stress direction is the slower of the two and it is the
faster if the stress is compressive.

The optical path difference R between the two waves when
they leave the glass is proportional to the product of the stress
σ and the thickness of glass traversed

$$R = C.\sigma.d$$

where C is a property of the glass, its stress optical coefficient. The value of C depends on the system of units used.
In the c.g.s. system the unit was called the Brewster. Fortunately the Brewster can be retained in the S.I. system, provided
that R is measured in nanometres, σ in MNm^{-2} and d in millimetres. The value of C does not vary greatly with composition

within the range of the most frequently encountered commercial silicate glasses. It is about 2.8 for container and flat glass compositions and between 3.0 and 3.5 for borosilicate glasses. For lead silicate glasses, the value of C decreases with increasing PbO content, eventually passing through zero and changing sign. The negative value of C means that the statements made above about the effects of tensile and compressive stresses on the velocities of the two components must be reversed.

Interference effects cannot be observed between plane polarised light waves unless the planes of polarisation coincide. To achieve this, the waves are passed through an analyser. This is usually a sheet of "Polaroid", a polarising filter having the property of transmitting only light vibrating in one particular direction. This analyser direction is set at right angles to the polarisation direction of the original plane polarised wave and hence at 45° to each of the resolved components (Fig. 47). An equal component of each wave is transmitted and, since these components are now vibrating in the same plane, interference effects are observed.

For the simple system described, the intensity of the light I transmitted by the analyser depends on the phase difference δ, between the two components. δ is related to R by

$$\delta = 2\pi R/\lambda ,$$

where both R and λ are measured in the same units. δ is in radians. The relationship between I and δ is

$$I = K\sin(2\theta).\sin^2(\delta/2)$$

and since we are considering only the situation when $\theta = 45°$, this simplifies to

$$I = K\sin^2(\delta/2) . \tag{41}$$

K is a constant proportional to the intensity of the original plane polarised wave.

Thus I varies periodically with δ. It is zero when $\delta = 0$, 2π, 4π etc. and has a maximum value of K when $\delta = \pi$, 3π, etc. These values of δ correspond to values of R of 0, λ, 2λ, etc. for zero intensity and of $\lambda/2$, $3\lambda/2$, $5\lambda/2$ etc. for maximum intensity. For the stresses normally encountered in glassware of a few millimetres thickness, R is usually less than one wavelength.

The retardation can be measured by compensation techniques.

Additional components are introduced into the light path, between the polariser and analyser by adjustment of which the retardation at a particular point in the field of view is reduced
to zero, usually by the compensator introducing a retardation
equal and opposite to that produced by the stress in the glass.
The compensators most commonly used are either of the Babinet-
Soleil quartz wedge or Berek rotating plate type (Partridge,
1949, Monack and Beeton, 1939). Other compensation techniques,
the so-called Senarmont and Tardy methods, do not use a compensator as such but give an optical effect very similar to that
of the Berek compensator. These compensators, or equivalent
equipment, are either supplied with a calibration or they can
easily be calibrated by the user.

For rapid inspection purposes, measurement of the retardation is not necessary. An approximate indication of the level
of stress is often all that is required. This can be obtained
using a strain viewer, which uses a polariser and analyser as
described above but white light from a tungsten lamp is used instead of a monochromatic light source. An extra component, a
tint plate, is introduced between the polariser and the analyser.
This is a very thin plate of a uniaxial crystal cut with its optic axis in the plane of the plate. It has an optical effect
similar to that produced by a piece of glass under uniaxial
stress and its thickness is chosen to give a uniform retardation
of approximately 550 nm over the whole field of view. On looking through the eyepiece of a strain viewer one sees a uniform
purple colour. The introduction of a piece of stressed glass
into the light path either adds to or subtracts from the tint
plate retardation producing other colours, the nature of which
gives information about the magnitude and nature of the stresses
present.

The explanation of why colours are produced depends partly
on the fact that white light has a wavelength range from about
400 to 750 nm and partly on the fact that the retardation R
(but not the phase difference δ) produced by a given stress is
almost independent of the wavelength of the light used. Suppose
for the moment that we are using a strain viewer with the tint
plate removed and that the stress in the glass is such that the
retardation is 300 nm. This is about $3\lambda/4$ for blue light and

about $\lambda/2$ for green light. Thus the δ values are different for the various components of white light. Equation 41 shows that the different colours will therefore be attenuated by the optical system by different amounts. The distribution of intensities in the white light will thus be changed and a colour will be seen. As the retardation increases the colour will change. The colours corresponding to various values of retardation up to 750 nm are listed in Table IX.

This shows that no definite colour appears until the retardation approaches 300 nm. Since, for thicknesses of glass normally encountered, satisfactory annealing should result in a retardation of less than 50 nm, a simple polariser and analyser arrangement is not sensitive enough. However, retardations produced by low stresses are readily detected once a tint plate is introduced. The total retardation due to the combined effect of the stressed glass and the tint plate can range between 500 and 600 nm. The colour changes in this range of retardation are easily seen. Table X gives the correspondence between colours and retardation for a strain viewer fitted with a tint plate.

Tensile and compressive stresses produce opposite effects on the total retardation. To identify the colour change corresponding to each type of stress the simplest procedure is to place a rectangular bar of annealed glass in the instrument and bend it in the vertical plane (assuming that the strain viewer is set up with the light path horizontal). The bent beam is then rotated slowly in the vertical plane until the brightest colours are seen. This will be the direction in which the directions of the compressive and tensile stresses are at 45° to the polarisation directions of the instrument. Note however that if the beam is rotated through 90° in the vertical plane, the colours corresponding respectively to tensile and compressive stresses will reverse.

The interpretation of the colours seen in a strain viewer is by no means simple. One needs some understanding of the stress pattern likely to be present in the sample to interpret what is seen with any confidence. More often than not one is looking not at a simple uniaxial stress system but at one which is much more complicated.

TABLE IX

Effect of retardation on the colour seen (tint plate removed)

Retardation nm	Colour
O	Black
50	Iron grey
200	Whitish grey
300	Yellow
425	Orange
530	Red
565	Purple
640	Blue
675	Blue green
740	Green
840	Yellow green
880	Yellow

TABLE X

Effect of retardation on the colour seen using a strain viewer equipped with a tint plate

Retardation introduced by the stress nm	Colour
+ 325	Yellow
+ 275	Yellow green
+ 200	Green
+ 145	Blue green
+ 115	Blue
O	Purple
− 25	Red
− 130	Orange
− 200	Bright yellow
− 260	Yellow
− 310	White
− 565	Black

For a biaxial system, or for a triaxial system in which one
of the principal stresses is directed along the light beam, the
retardation is proportional to the algebraic difference between
two principal stresses which are at right angles to the light
beam. Calling these stresses σ_x and σ_y, the retardation is
given by

$$R = C(\sigma_x - \sigma_y)d \ . \tag{42}$$

In using this equation, the stress values must be given the
appropriate signs, i.e. tensile stresses are taken as positive
and compressive stresses are negative. Thus if σ_x and σ_y are
equal and of the same sign, the retardation will be zero.
Another situation which can result in zero retardation is when
one is looking through a balanced stress system, i.e. one in
which the stresses change from tensile to compressive along the
light path. Two simple examples have been considered in previous sections: the sandwich seal and a uniformly cooled sheet.
In both cases the integral $\int \sigma \, dx$ is zero along a direction normal to the sheet surfaces. Because of this, the retardation is
also zero when polarised light is passed through the glass in
this direction.

Figure 48 shows the quite complex strain colour pattern

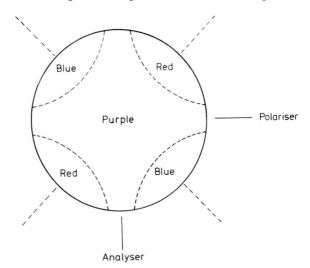

Fig. 48. Appearance of an imperfectly annealed disc when
examined in a strain viewer.

which one sees in a disc which has been cooled from above the annealing point. In such a disc there are stresses parallel to the surface which do not vary very much from the centre to the edge. These are due to the temperature gradients normal to the plane surfaces. With the light passing normal to the plane surfaces, these stresses will produce no retardation for the reasons just given. However, during cooling there will also be temperature gradients in a radial direction. These produce a radially symmetrical stress distribution. At any point, the principal stresses are a radial stress σ_r and a circumferential stress σ_θ. Together they produce a retardation proportional to $(\sigma_r - \sigma_\theta)$ provided that the two stresses make an angle at the point in question with the polarisation direction of the instrument. The purple cross seen in the pattern defines the two directions for which σ_r and σ_θ are parallel and perpendicular to the polarisation directions. Maximum colours are seen in the alternate 45° quadrants. Theoretical analysis of the stress distribution produced by cooling a disc from above the annealing temperature shows that at the centre of the disc σ_r and σ_θ are equal and have the same sign. Thus in the regions near the centre, retardations are low irrespective of the orientation of the stresses. At the disc edges, simple considerations of stresses at free boundaries lead to the conclusion that σ_r is zero. Thus at the edge, the colour is due entirely to the tangential stress which is compressive. The alternation of colours in the four quadrants of the colour pattern is due to the fact that in the NE and SW quadrants the tangential stress σ_θ at the edge is at right angles to its direction in the NW and SE quadrants. The reader may at this point suspect that this example has been perversely chosen for its complexity. This is only partly true. One has only to look through the base of an imperfectly annealed milk bottle to appreciate its practical relevance.

CHAPTER IV

THE STRENGTH OF GLASS

It is difficult to think of more than one or two applications of glass in which it is used primarily as a structural
load-bearing material.   The only one which comes readily to
mind is the use of glass fibres for reinforcing plastics and,
more recently, cement.   Normally its use depends primarily on
some other property, such as its transparency or high electrical
resistivity.   However, in many applications the glass is subjected to stress and it is then essential to design the glassware with a good understanding of the factors affecting its
strength.   If this is not done, the consequences may be very
serious.   Large windows must be thick enough to stand the maximum expected wind pressures and bottles containing aerated
drinks must not explode.   Even the implosion of a T.V. tube
might be considered a very serious matter, depending on the nature of the programme being shown at the time.

Glasses are brittle materials (with the exception of some
metallic glasses) and like other brittle materials, such as cast
iron, are weak in tension.   Typical fracture stresses for commercial glassware are in the range 50-150 MN m^{-2}, which may be
compared with 3000 MN m^{-2} for some high strength steels.   An
engineer having to use a brittle material in a load-bearing component will always try to ensure that the stresses in the component are compressive.   This practice was universally followed
in designing the first iron bridges, e.g. the famous example,
still standing in Coalbrookdale, Shropshire.   When glasses are
used in this way they are immensely strong and may be preferred
to metals.   A remarkable example of this is the proposed design
by the U.S. Naval Ordnance Laboratory of components for a deep-
diving submarine (Ernsberger, 1965) in which the crew, the motor
and certain critical electrical components are housed in three
spherical pressure hulls made from glass, held together in a
tubular frame structure.   Under water the glass is subjected
to a uniform hydrostatic pressure.   Provided that entry hatches
and electrical feed-throughs in the walls of the hulls are carefully designed, it is possible to dive to great depths without

any risk of them fracturing or collapsing.

When a glass article breaks, the fracture nearly always originates from a single point situated somewhere on the surface of the article. It is unusual to find a fracture starting from an internal point, unless the glass is of so poor a quality that it contains foreign inclusions. If the stress in the glass is low when it fractures, the crack may propagate quite slowly. It is not uncommon to see cracks, starting from the edge of a window pane, which take weeks to go from one edge of the pane to another. The main characteristics of a fracture formed when the stress is high is that the cracks move at high speed - the article appears to fragment instantaneously and the crack pattern shows much forking (Fig. 49).

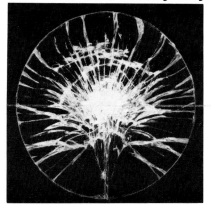 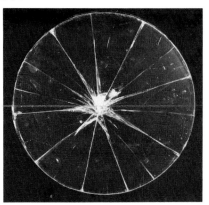

Fig. 49. Fractures in discs broken by bending. The fracture origin is at the edge of the disc on the left and is at the centre of that on the right.

Another important general feature of the tensile strength of glasses is that values measured on a sample of apparently identical specimens, or items of glassware made at the same time and apparently treated in the same way, nearly always show a considerable variation. A coefficient of variation of 10-15% is not unusual. This is not peculiar to glasses. It is a general characteristic of all brittle materials.

In what follows in this chapter, we shall assume that we are discussing the strength of glass which is of reasonably good commercial quality. Badly melted glass will contain chemical inhomogeneities, as a result of which the thermal expansion coefficient will vary considerably from point to point. Internal stresses will be produced as the glass is cooled to room temperature and these may seriously reduce its strength.

A.   Theoretical strength of glass

There are a number of methods whereby one can estimate the strength of a material from the strength of the interatomic binding forces.   One such, due to Orowan (1949), estimates the cleavage strength of a brittle material.   Good accounts are available in a number of text books (Kelly, 1973;   Cottrell, 1964), and the details of the theory will therefore not be repeated here.   The physical idea used in the theory is that mechanical work is done in stretching to the breaking point the bonds acting across a plane in the material at right angles to the stress direction.   This work is equated to the "fracture surface energy" of new surfaces created by the fracture.

This quantity, the fracture surface energy, γ (units of Joule m^{-2}) is very important in the study of brittle fracture. The energy of the newly-formed fracture surface arises because the forces acting on the surface atoms are primarily inward-acting forces of attraction from atoms in the interior, whereas atoms in the interior are subjected to forces which are practically the same in all directions.   The surface tension of liquids may be explained in similar terms.   Until recently γ could only be estimated, but now there are many methods by which it can be measured.   Wiederhorn (1966) has found that, for a commercial soda-lime-silica glass, the value is in the range 2-4 Jm^{-2}, according to the ambient conditions.   Also it is relatively insensitive to glass composition (Table XI).

TABLE XI

Fracture Surface Energies (Wiederhorn, 1966)

	γ Jm^{-2}	
	in N_2 at 25°C	in liquid N_2 at -196°C
Vitreous silica	4.32	4.56
Soda-lime-silica	3.82	4.53
High lead oxide content silicate	3.50	4.11

The Orowan theory leads to the following equation for the
theoretical fracture stress

$$\sigma_m = (E\gamma/a_o)^{\frac{1}{2}} \qquad (43)$$

E is the Young's Modulus of the material and a_o is the interatomic distance.

For many commercial glasses, E is approximately 10^{11} Nm^{-2}
and a_o is of the order 2 x 10^{-10} m. Using a value of 3 JM^{-2}
for γ, one finds that σ_m is 4 x 10^4 MNm^{-2}, a value very much
higher than is normally found in practice.

It is worth noting that, for the range of commercial oxide
glass compositions, none of the terms in the Orowan equation is
much affected by changes in glass composition. Thus the theoretical strength of glass is about the same for all glasses of
practical interest.

B. Effect of surface flaws on the strength of glass - the Griffith equation

The fact that measured fracture stresses are normally much
less than the theoretical value is explained in terms of the
presence of submicroscopic flaws in the glass surface. Such
flaws act as stress concentrators, the stress at the flaw tip
being much greater than if there were no flaw present. The
actual geometry of the flaws in a glass surface may be complicated, and at present we have no means of studying flaw shapes.
So, by necessity, one is obliged to consider the likely effect
of variations in flaw size and shape in terms of simple
geometries.

For the "flaw" represented by a narrow elliptical hole in
a plate shown in Fig. 50, Inglis (1913) has shown that the stress
at the flaw tip σ_{yy} is given by

$$\sigma_{yy} = \sigma(1 + 2c/b) \ . \qquad (44)$$

Since the radius of curvature, ρ, at the end of the major axis
of an ellipse is equal to b^2/c

$$\sigma_{yy} = \sigma(1 + 2\sqrt{c/\rho}) \qquad (45)$$

and when c is large compared with ρ

$$\sigma_{yy} \approx 2\sigma\sqrt{c/\rho} \qquad (46)$$

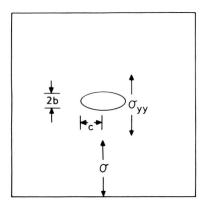

Fig. 50. Elliptical flaw in a glass plate under tension.

One would expect that if σ_{yy} were to exceed the theoretical fracture stress of the material, the flaw would extend and the material would fail.   This is not necessarily so.   A.A. Griffith (1920) made a very important theoretical contribution to the general understanding of brittle fracture by demonstrating that for a given value of the applied stress, σ, the flaw will extend and fracture occur only if the crack has a length greater than a certain value, and that this critical crack length is determined by *energy* considerations.   Griffith's theory is well explained in a text book by Cottrell (1964).   His account of the Griffith theory, which has the merit of being easy to follow, will be reproduced here.

For a plate of unit thickness which is subjected to a uniform stress, σ, the strain energy per unit volume is

$$\sigma e/2 = \sigma^2/2E$$

where e is the tensile strain produced by the stress.   Suppose such a plate is stretched and then held between two rigid clamps as shown in Fig. 51.   Now consider the change in the energy of the plate which will result from the introduction of a crack of length 2c running through the thickness of the plate. Obviously the stresses in the plate will be greatly modified in the immediate vicinity of the crack.   Near the crack tips the stresses will be greater than σ but the stress component normal to the crack surface will be zero.   At distances from the crack which are large relative to the crack dimensions, the stress will have

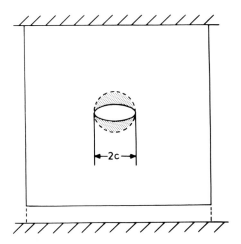

Fig. 51. Elliptical flaw in a glass plate under tension.

the same value as that before the crack was introduced.   Although the stress distribution around the crack can be calculated
exactly, Cottrell suggests that, as a first approximation, it is
assumed that, within a circle of radius c, the centre of which
coincides with the centre of the crack, the stress is zero, but
outside that circle it remains at the original value before the
crack was introduced.

The introduction of the crack thus *reduces* the strain energy
of the plate by an amount $\sigma^2 . \pi c^2 / 2E$.   However, the presence of
the crack increases the total surface area of material by an
amount of approximately 4c and thus there will be an *increase*
of surface energy of $4c\gamma$.

The net change in energy ΔE is given by

$$\Delta E = - \sigma^2 . \pi c^2 / 2E + 4c\gamma \ . \tag{47}$$

When c is very small, the first term is less than the second
and ΔE is positive, but for large values of c the first term
dominates and ΔE decreases continuously as c increases (Fig. 52).
Under the latter circumstance, the crack is energetically unstable and should grow spontaneously.   The critical condition
which determines the fracture stress is obtained by differentiating Equation (47) with respect to c and setting the result
equal to zero.   This gives c*, the value of c at which the
crack becomes unstable at the applied stress σ_f.   This is

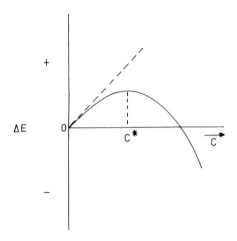

Fig. 52.   Change in energy as a function of crack length.

$$c^* = 4\gamma E/\pi\sigma_f^2 . \qquad (48)$$

An exact analysis of the change in strain energy due to the introduction of an elliptical crack of length c gives a value which
is exactly twice that given above.   Hence

$$c^* = 2\gamma E/\pi\sigma_f^2 . \qquad (49)$$

Considering the problem the other way round by asking the
question "What is the value of the stress at which a crack of
length c^* will become energetically unstable?", rearrangement of
Equation (49) gives

$$\sigma_f = \sqrt{2E\gamma/\pi c^*} \qquad (50)$$

Cottrell shows that exactly the same result is obtained
when one considers the more "natural" situation, in which the
plate is not clamped but is allowed to stretch as the crack
grows.

Other authors, who have analysed problems involving cracks
with different but still simple geometries, have derived formulae similar to Equation (50), apart from the numerical factor
$2/\pi$.   However, the value of this factor is not greatly affected
by crack shape.   For a fracture stress of 100 MN m^{-2}, typical
of commercial glassware and values of E and γ given in the previous section, one finds that the critical flaw size c^* is
2×10^{-2} mm.

Since the Griffith equation is based on an energy criter-

ion, it is a necessary, but not a sufficient condition for
crack growth.   One has also to consider whether or not the
stress at the fracture tip exceeds the theoretical fracture
stress.   If it does not, then the crack will not extend even
though the energy condition is satisfied.   However Orowan
(1955) in discussing this point has argued that in a truly
brittle solid, the effective value of the radius ρ at the crack
tip will be of the order of the interatomic spacing in the material.   A consequence of this is that if the applied stress is
high enough to satisfy the Griffith criterion the stress at the
crack tip will also be high enough to exceed the theoretical
fracture stress (Kelly, 1973).

The early work of Griffith and Inglis has been greatly extended in recent years and their ideas have been applied to the
analysis of fracture in materials other than glass.   The main
impetus for this work has been the need to design large metal
structures in such a way that there is no possibility of their
failing in a sudden catastrophic manner by brittle fracture.
This modern study of the process of fracture is known as
"Fracture Mechanics" and it is appropriate here to explain the
meaning of a term which has been introduced in the development
of this science, and which is now widely used in the literature.
This is the stress intensity factor, K.   As its name suggests,
it is a measure of the extent by which a crack magnifies the
level of stress in the immediate vicinity of its tip.   It is
a function of the polar co-ordinates, r and θ, (Fig. 53) of
the point at which the stresses are calculated with respect to
the crack tip as origin.   The form of the function depends on
the way in which the material containing the crack is stressed

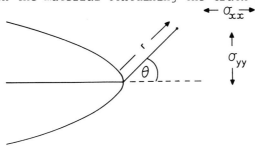

Fig. 53. Stresses near a crack tip.

and there are three such functions denoted by K_I, K_{II} and K_{III}.
Only K_I will concern us.   This is the function which applies
when the material is subjected to a tensile stress in the direction at right angles to the crack direction.   For a plate of
infinite extent containing a through crack of length 2c, K_I is
given by

$$K_I = \sigma \sqrt{\pi c} .$$
   (51)

and the stresses σ_{xx} and σ_{yy} close to the crack tip are

$$\sigma_{xx} = \sigma (\pi c/2\pi r)^{\frac{1}{2}}.f_1(\theta) = K_I.f_1(\theta)/(2\pi r)^{\frac{1}{2}}$$
   (52)

$$\sigma_{yy} = \sigma (\pi c/2\pi r)^{\frac{1}{2}}.f_2(\theta) = K_I.f_2(\theta)/(2\pi r)^{\frac{1}{2}} .$$

where $f_1(\theta)$ and $f_2(\theta)$ are functions of the polar angle, θ.   As
the applied stress increases, the value of K_I increases and
fracture eventually occurs at the fracture stress σ_f when

$$K_I = \sigma_f (\pi c)^{\frac{1}{2}} .$$
   (53)

This value of K_I is termed the critical stress intensity factor
K_{IC}.   It is related through the Griffith equation to the fracture surface energy γ by

$$K_{IC} = (2E\gamma)^{\frac{1}{2}}$$

K_{IC} is determined experimentally using a test piece of simple
geometry in which a macroscopic crack of length c is made before
the test is carried out.   The value of K_I corresponding to the
applied stress σ for a specimen of finite dimensions is calculated from the equation

$$K_I = \sigma Y c^{\frac{1}{2}}$$
   (54)

where Y is a dimensionless factor which depends on the geometry
of the crack and on how the loads are applied to the test piece.
Equations for calculating Y have been derived for many types of
crack, specimen geometry and method of loading.

    The experimental determination of K_{IC} amounts to a measure-
ment of the fracture surface energy, which is a more fundamental
measurement of the property of the material than the measurement
of the fracture stress.   Any discussion of the fracture process
is best carried out in terms of the stress at the crack tip,
measured by K_I, rather than in terms of the fracture stress,
which is merely an average stress throughout the bulk of the
material.

C. Methods for measuring glass strength

When measuring the strength of glass and glassware, one is faced with a situation which is quite different from that which exists in the metallurgical field. Strength measurements made on metal specimens in the laboratory provide information which can be used in the design of large metal structures. It will be clear by the end of this chapter that the strength of a glass specimen or article is largely determined by the number and severity of the flaws in its surface. Strength measurements made on glass rods, for example, are of no value in predicting whether or not a glass container will withstand a certain internal pressure, even though the rods and the container are made from the same glass. Because the rod and the container are made by radically different manufacturing processes and will have been subjected to different kinds of damage prior to testing, the strength properties of the surfaces of the rod and of the container will usually be quite different.

However, a great deal of laboratory work is done on the strength of glass using specimens of simple geometry, the main aim of this work usually being to study in detail those factors which affect the strength of glasses as materials. From work of this type, a much clearer understanding has been gained of the factors affecting the strength of manufactured glassware, even though, for the reasons, explained, the laboratory results may not be directly useful for design purposes.

The most readily available forms of glass for laboratory studies of strength are cylindrical rod, flat glass and fibre-glass. In the following paragraphs a brief account will be given of some of the methods which have been used for measuring the strength of samples in these forms.

1. Strength measurements on rods

It is difficult to measure the strength of rods by the application of axial tension. Unless great care is taken in mounting the specimen in the grips of the tensile testing machine, unknown bending forces will be added to the measured tensile forces. Also, fractures frequently occur near the grips either because of local stress concentrations or because the grips have caused further damage to the glass surface.

For these reasons the strength of rod specimens is usually
measured by three- or four-point bending (Fig. 54).   The bend-

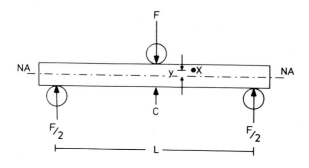

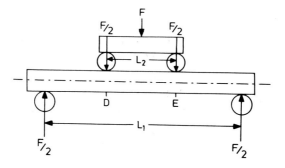

Fig. 54.   Three- and four-point bending.

ing forces produce stresses acting parallel to the rod axis
which are tensile below the horizontal plane NA (the neutral
axis) and compressive above it.   The magnitude of the axial
stress at a point such as X is given by simple beam theory

$$\sigma = My/I \qquad (55)$$

y is the vertical distance of X from plane NA, **M** is the bending
moment at X due to the applied forces and, for rods of constant
circular cross section,

$$I = \pi R^{b}/4 \ . \qquad (56)$$

For three-point bending, the maximum tensile stress is at C
immediately below the central knife edge.   At C, y = R and
M = FL/4.   Hence

$$\sigma_{max} = FL/\pi R^3 \ . \qquad (57)$$

For four-point bending, the maximum tensile stress is con-

stant along a line in the specimen running from D to E.   For
all points on the line y = R and for all planes between D and
E, M is given by

$$M = F(L_1-L_2)/4 \qquad\qquad (58)$$

Thus for four-point bending

$$\sigma_{max} = F(L_1-L_2)/\pi R^3 \qquad\qquad (59)$$

In both equations, if F is in Newtons and L and R in metres, the
stress values will be in Newton m^{-2}.

In measuring the strength, the forces are increased at a
constant, controlled rate until fracture occurs.   The maximum
stress in the rod when fracture occurs is calculated.   This
stress is called the Modulus of Rupture.   The true fracture
stress will usually be lower than this.   It is very unlikely
that, in any bending strength measurement, the most severe flaw
in the specimen will be where the stress is greatest, i.e. at C
in three-point bending or on line DE in four-point bending.
Fracture normally occurs at a point where there is a severe flaw,
but not necessarily the most severe, and where there is a high
stress, but not necessarily the highest stress.

It has been suggested that the four-point loading method is
preferable to three-point loading, in that the stressing of the
specimen is more uniform and might be expected to result in a
reduced variability in the Modulus of Rupture values.   However
there does not seem to be much evidence to support this opinion.
Whichever method is used, it is desirable that the nominal dimensions of the rods used, the spacing of the loading points and
the rate of loading should be specified when reporting results.

2. Strength measurement on flat glass

A method which has been fairly widely used on sheet glass
is to determine the Modulus of Rupture in either three- or four-
point bending of strips cut from the sheet.   The formulae for
calculating the stresses are similar to those given in the previous section, but the value of I is now $wt^3/12$; w is the width
of the strip and t is its thickness.   The maximum tensile
stresses occur on the lower surface of the strip.   The value
of y in Equation (55) is therefore t/2.

When a four-point bending apparatus is used, the stress is

uniform over the lower surface of the strip between the two
loading knife edges.   Since the fracture nearly always occurs
in this region, the Modulus of Rupture and the fracture stress
are equal.   The method is however open to criticism in that
the process of cutting the strips from the sheet produces flaws
at the cut edges of the specimen, and the fracture may originate
from one of these flaws.   The results will therefore be misleading if their purpose is to measure the strength of the parent sheet.   The difficulty can be overcome if transparent adhesive tape is applied to the compression face of each strip before it is broken.   The fragments then hold together and it is
easy to identify which fractures originate from the sheet edge
(Kerper and Scuderi, 1964).   If it is not the purpose to measure the strength of the parent sheet, one can apply a controlled abrasion treatment, e.g. by sand blasting, to a circular
area in the centre of the tension face of the strip.   Mould and
Southwick (1959) used this method in an extensive study of the
effect of time under load on the strength of abraded glass
surfaces.

Another method of measuring the strength of flat glass,
which reduces the probability of fracture at the cut edge, is
to break circular discs by applying bending forces.   The various
methods of loading are illustrated in Fig. 55.   All produce
tensile stresses at the edge of the disc, but these are less
than the maximum tensile stress, which in all cases is at the
centre of the disc on its lower surface.   In the concentric
ring loading method (Fig. 55b), the stress is constant within
the circular area bounded by the loading ring.   It is important
that the glass should be flat so that it makes a uniform contact with the support ring and the loading ring, if one is used.
For this reason, the method is only suitable for testing material such as ground and polished plate glass or float glass.
Even for these materials, it is advisable to use gaskets between the support and loading rings and the glass to help obtain uniform loading.

Finally, an interesting method which has been used on thick
plate and float glass is one in which a ball, usually made of
steel, is pressed against the glass by a force, F, which is
gradually increased until a fracture is seen to form around the

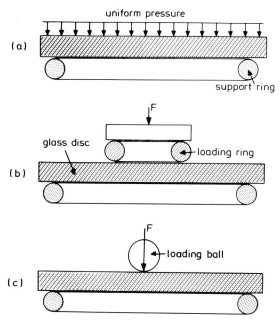

Fig. 55.   Methods of measuring the bending strength of glass discs.   (a) Uniform pressure loading. (b) Concentric ring loading.   (c) Central loading.

circular area of contact produced as the ball and glass deform. Figure 56 is a vertical section through the centre of the ball,

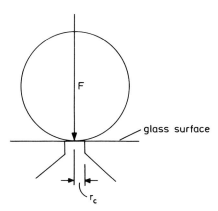

Fig. 56.   Hertzian fracture.

and shows that the crack, in the region close to the glass surface, is cylindrical with radius, r_c, which is usually somewhat

greater than the radius of the circule of contact, a.   Further
into the glass, the crack spreads out as a cone.   Cracks like
this are formed when a small high speed missile, e.g. a shot gun
pellet, strikes a window pane.

The theory of the stress distribution produced in this way
was worked out by Hertz (Timoshenko and Goodier, 1951) and, for
this reason, the typical cone fracture produced is usually called
a Hertzian fracture.   At the glass surface, the maximum tensile
stress is produced at the edge of the circle of contact, i.e.
at a distance r = a from the centre of this circular area.   The
stress acts in a radial direction, and is given by the equation

$$\sigma_a = K_1 F/a^2 \tag{60}$$

where K_1 is a constant, the value of which depends on the elastic
properties of the glass.   The radius of the contact circle, a,
is given by

$$a = (K_2 FR)^{1/3} \tag{61}$$

the value of K_2 depending on the elastic properties of both the
glass and the ball.   Combining the last two equations one obtains

$$\sigma_a = K_3 (F/R^2)^{1/3} . \tag{62}$$

As stated earlier, the radius, r_c, of the circular crack at the
glass surface is invariably somewhat greater than a, and consequently the fracture stress, σ_f, is less than σ_a. σ_f can be
calculated if r_c is measured, from

$$\sigma_f/\sigma_a = (r_c/a)^2 . \tag{63}$$

The method has the advantage that, because the cracks are small
and usually do not penetrate far into the glass plate, many
measurements can be made on a single specimen of plate.   However, we shall see later that fracture stresses measured by this
method are usually significantly greater than the value obtained
when the strength is measured by a bending method.   The area of
glass surface stressed in a Hertz fracture measurement is very
small and, because the larger flaws which determine the strength
when large areas are stressed are rather far apart, one rarely
locates such a flaw within the highly stressed region round
the edge of the contact circle.

In most investigations using the Hertz method, steel ball bearings have been used. However Johnson et al. (1973) have shown that, if the ball has a different elasticity from the glass, additional stresses are produced by friction as the deforming surfaces are brought into contact. These stresses are not taken into account in the Hertz theory. To avoid this difficulty, glass balls should be used.

3. Strength measurement on fibres

Glass fibres commonly have strengths much greater than bulk glass. Their extensive use in glass-reinforced plastics is one reason why many studies have been made of factors affecting the strength of fibres. Strength measurements are usually made by tensile testing. Special features of technique are the methods used for obtaining the sample of fibre and for transferring it to and mounting it in the testing machine. Useful details of the apparatus and techniques may be found in papers by Thomas (1960) and Metcalfe and Schmitz (1964).

D. High strength glass surfaces

Very strong glass specimens can be made in a number of ways, either by forming the glass from the melt in such a way that no damage to the surface can take place whilst it is cooled and transferred to the testing apparatus, or by removing the damaged surface layer from specimens which have already suffered some loss in strength due to handling.

Cornelissen and Zijlstra (1962) produced rods and fibres of soda-lime-silica glass with strengths up to 3000 MNm^{-2}. Two methods were used, by drawing from the surface of glass in a crucible and by hand drawing after heating in a flame. The highest strengths were obtained at the highest drawing temperatures. Thus, drawing after heating in a gas-oxygen burner to about 1000°C gave strengths of 2750 MNm^{-2}. After heating to only 800°C a lower strength of 300 MNm^{-2} was obtained. High values of approximately 3000 MNm^{-2} were also measured by Thomas (1960) on fibres of E glass drawn from an electrically heated platinum bushing. The temperature of the drawing orifice was in the range 1220-1340°C. If silica glass is strongly heated in an oxy-hydrogen flame, the surface material volatilises.

By this method Hillig (1961) produced 0.5 mm diameter fibres
with strengths measured at 78°K, as high as 14,000 MNm^{-2}.

The damaged surface can also be removed by acid etching.
Figure 57 shows results of Proctor (1962) on soda-lime-silica

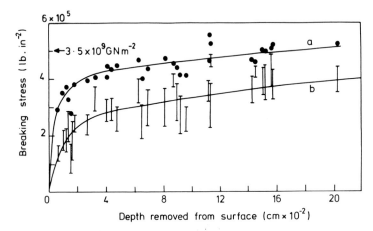

Fig. 57. Effect of acid etching on the strength of glass rods.

glass using an etchant consisting of an aqueous 15% HF, 15%
H_2SO_4 solution, in which the glass rods were agitated at a temperature of 35-40°C.   Unfortunately, no means has yet been discovered of completely protecting these extremely strong surfaces.
Although it is possible to minimise damage due to handling, the
greater proportion of the high strength is soon lost, even if
the glass is treated with the most effective surface protection
process suitable for general use.   (These statements may soon
require qualification in the light of recent work on the application of certain polymer coatings to freshly drawn silica
fibres for communication applications.)

E.   Damage to glass surfaces

1. Mechanical and thermal damage

The high strength, pristine surfaces produced by the methods
just described may suffer mechanical damage by impact or by sliding contact with some other hard surface.   Loss of strength may
also result from heat treatment of the glass.

Even what may appear to be quite gentle handling reduces

the strength drastically.   Thus Zijlstra (1962) found that rods
having an initial strength of 2000 MNm^{-2} suffered a strength
loss of 20-40% when their surfaces were rubbed with paper. Roll-
ing one against another reduced the strength to 400 MNm^{-2}, and
when one rod was rubbed vigorously against another the strength
fell to a value typical of commercial glassware, between 100 and
150 MNm^{-2}.   To reduce the strength further, rather severe treatment is needed.   Thus sandblasting may reduce the strength to
about 15 MNm^{-2}.

One reason why it is so easy to damage a very clean glass
surface is that the coefficient of friction between two such
surfaces is very high - about 1.   High stresses are produced
when one surface is slid against another.   The sliding is very
jerky and there is considerable wear of the surfaces where they
have been in contact.   Damage of this kind can be greatly reduced if a lubricating coating is applied to the glass, and these
are now widely used in the manufacture of glass containers. The
coating is usually applied by spraying a solution or emulsion
of the coating material onto the glass as it leaves the cold end
of the annealing lehr.   The coatings are most effective if they
are also resistant to wear.   A high enough pressure applied between two coated surfaces will result in rupture of the lubricant film, so that one has again direct glass-to-glass contact
and a high coefficient of friction.   It has been found that the
application of a very thin coating of either SnO_2 or TiO_2 to
the surface of the hot container, shortly after the glass is
released from the blow mould of the container machine, produces
a surface which confers enhanced wear resistance (a higher breakdown load) on the lubricant film when this is subsequently applied.   The "hot end" oxide coating is usually applied by passing
ing the hot containers on a conveyor belt through a chamber into which is fed an air stream containing a controlled concentration of a volatile tin or titanium compound.   Tin tetrachloride is commonly used on account of its cheapness. Gaiser
et al. (1965) and Zinngl and Simminsköld (1967) have presented
results demonstrating the effectiveness of these coatings when
applied to containers.   Dettre and Johnson (1969) and Turton
and Rawson (1973) have carried out laboratory investigations
using microscope slides coated with various protective coatings,

with the object of understanding why they are effective.

The other way in which high strength surfaces may be damaged is by heat treatment. Cornelissen and Zijlstra (1962) found that the strength of pristine rods was reduced to 400 MNm^{-2} after heating in air at 630°C for 30 min. and to 200 MNm^{-2} after heating at 850°C for 2 min. They attributed the strength reduction to slight devitrification of the glass during heat treatment. One would expect that if the temperature of heat treatment had been increased further to, say, 1000°C, no decrease in strength would have occurred since this temperature would probably have been above the liquidus temperature of the glass used.

Brearley and Holloway (1963) have observed similar effects. They also found that the strength is reduced by heat treatment even at temperatures well below the transformation range. It is very unlikely that devitrification would occur at such low temperatures. Their observations showed that a further source of damage was connected with bonding of dust particles from the atmosphere to the glass surface.

These observations are of some practical relevance to the strength of hollow articles such as glass containers and lamp bulbs. Their strength is usually determined by the strength of the outer surface, which is considerably less than that of the inner surface. However if the articles are subjected to certain types of loading, e.g. squeezing between parallel plates or to impact, the inner surface is highly stressed and fracture may occur from that surface if it is weak. It has been shown that the impact strengths of glass containers made by the press and blow process are less than those of similar containers made by the blow and blow process. This lower strength may be correlated with the presence of a greater concentration of minute, contaminating particles which can be observed on careful observation of the internal surfaces of containers. Also Zijlstra and de Groot (1962) have shown that the crushing strength of 60 mm diameter lamp bulbs made from glass gobs is higher than that of bulbs of similar dimensions made by the flame working of glass tubing; i.e. the higher the temperature of forming, the greater the strength.

2. Revealing the extent of damage on a glass surface (The Ernsberger technique)

The interpretations of the experiments on damage mechanisms just described are well supported by observations made using a technique devised by Ernsberger (1960, 1962), which gives a direct visual indication of the state of damage. The glass is immersed for a controlled period of time, 30 min. say, in a fused eutectic mixture of $LiNO_3$ and KNO_3 (60 mol% KNO_3). The most suitable temperature has to be determined by experiment for the particular glass being used. For a container glass, a temperature in the range 220°-250° will normally give satisfactory results. Base exchange occurs between the salt bath and the glass. Sodium ions leaving the glass are replaced by an equal number of lithium ions from the salt. The lithium ions are significantly smaller than the sodium ions (Li^+ 0.6 A.U., Na^+ 0.95 A.U.). As a result, the thin surface layer which has been affected by the base exchange attempts to shrink, but is restrained from doing so by the interior of the glass, the composition of which is unaffected. Consequently, a tensile stress is produced which has the same value in all directions parallel to the surface. The magnitude of the stress is determined by the temperature and time of the base exchange process. If it is high enough, some of the flaws act as fracture origins, and surface cracks are formed running from these flaws. After the specimen has cooled and has been washed free from salts, the cracks may be made more visible by a short etch in dilute hydrofluoric acid. Some preliminary experiments are usually needed to obtain the best results. It is best to use a fixed base exchange time and then determine by trial a temperature which will clearly reveal only the more severe flaws. If too high a temperature is used, the crack pattern may be so dense that it is difficult to interpret. Also the cracks begin to propagate parallel to the surface, and, finally, flakes of glass break away from the surface.

Ernsberger's papers contain many excellent and interesting photographs, a few of which are reproduced here. Figures 58 and 59 show mechanical damage, Figs. 60 and 61 damage produced by heat treatment.

118

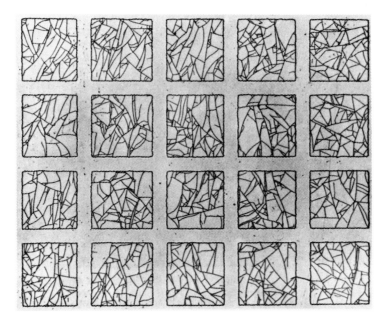

Fig. 58. Damage produced by impact. Part of the surface was masked by a resist so that the flaws appear only in the unmasked areas, which are 0.5 cm square.

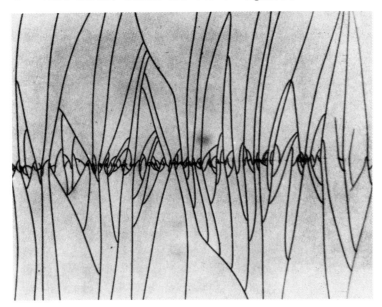

Fig. 59. Damage produced by sliding the tip of a glass fibre across the surface (x 20).

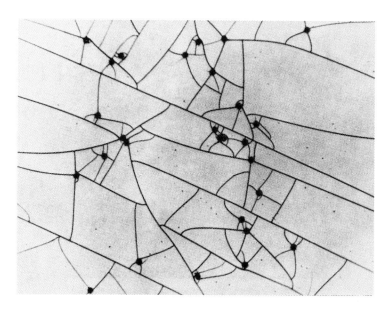

Fig. 60. Previously etched surface after heating sheet glass
for 2 hr at 650°C (x 75).

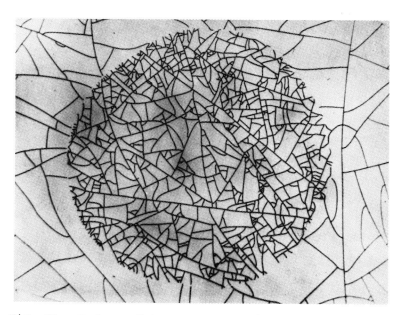

Fig. 61. A drop of tap water was allowed to evaporate on the
surface, after which the glass was heated for 1 hr at 650°C
(x 20).

Ernsberger suggests that the dark spots on the surface in
Fig. 60 are minute crystals resulting from devitrification of
the glass.   Finally, Fig. 62 is from the author's laboratory.
It shows a patch of severe damage on the internal surface of a
glass container.   This was produced during the manufacture of
the container and is associated with the contamination of the
glass surface by minute solid particles, which are either metallic in nature or are residues from lubricants applied to the
plunger which preforms the internal surface of the container.

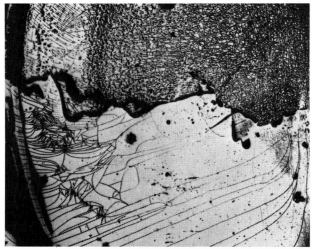

Fig. 62.   Damage on the external and internal surfaces of the
base of a milk bottle.   The internal surface is on the left.

F.   Factors affecting the strength of damaged surfaces

Usually, when one is measuring the strength of glass specimens or articles of glassware, the measurements are made on a
representative sample in which the individuals have had a similar history and which have been damaged to a similar extent.
In this section we consider some of the factors, other than the
degree of damage, which affect the results obtained.

1.   Area under stress

A feature of the type of damage normally encountered on the
surface of glass articles is that the most severe flaws, i.e.
those which determine the strength of the weakest individuals in
the sample, are relatively few in number.   Consequently, one

finds that the smaller the area of glass surface which is subjected to stress, the higher is the fracture stress, and also
that the fracture stress depends on how the method of loading
used in the strength measurement distributes the stress over the
glass surface.   In the following, some examples are given to
illustrate these points.

Table XII contains Modulus of Rupture results obtained by
Kerper and Scuderi (1964) on 1½" wide strips of plate glass.
One group of samples was tested with three-point loading and the
other three groups with four-point loading.   For each group of
50 specimens, the loading points were set at different distances
apart (Fig. 54).   The position of the fracture origin was located on every specimen, so it was possible to calculate the
fracture stress as well as the modulus of rupture.   Specimens
breaking from edge flaws were not taken into account.

TABLE XII

Effect of area under maximum stress on the modulus of rupture
of strips tested in three- and four-point bending

ℓ_2 mm	Area under maximum stress sq. mm	Modulus of Rupture MNm^{-2}	Stress at Fracture MNm^{-2}
O		138	124
50	1900	126	121
100	3800	110	108
200	7600	96	94

The second set of results were obtained by Metcalf and
Schmitz (1964) in a study of the effect of specimen length on
the strength of fibres of E glass of 10 μm diameter.   Figure 63
shows that the magnitude of the length effect depends very much
on the state of damage.   There is a marked effect for the fibres
labelled ES.   Each fibre was obtained by separating it from a
commercially produced strand of 65 parallel fibres.   The EM
fibres, for which the length effect is much smaller, were obtained from fibre drawn as a monofilament, a procedure which
results in far less damage.

Finally Table XIII gives fracture stresses of float glass

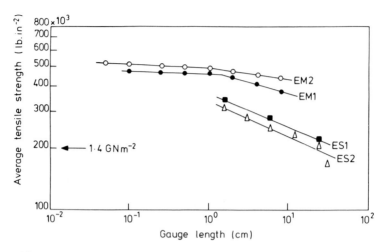

Fig. 63.   The effect of length on the strength of glass fibres.

TABLE XIII

Effect of ball diameter on the fracture stress measured by the
Hertz method on the two surfaces of float glass

Ball diameter mm	Mean Fracture Stress	
	Tin bath surface	Upper surface
	MNm^{-2}	
6.35	350	600
3.17	480	670
1.59	620	750
0.79	800	900

surfaces measured by Hamilton and Rawson (1970) using the Hertz
fracture method.   The decrease in strength with increasing ball
size is due to the increasing area of contact between the ball
and the glass and hence the increasing area which is subjected
to high stress.

These results show a number of interesting features.   The
surface which has been in contact with the tin bath is the weaker
of the two.   This is the surface which is subjected to damage
as it passes over the supporting rollers in the annealing furnace
during the manufacturing process.   The strength difference is

not so obvious with the smallest ball used, presumably because
this stresses such a small area that the more severe flaws produced during the manufacturing process are rarely detected.
Even with the largest ball size used, the strength is two or
three times greater than fracture stresses measured on the same
glass in plate bending experiments.

2.   Flaw statistics

These area effects can be analysed mathematically using the
methods of flaw statistics.   From such an analysis, one can
obtain a quantitative description of the state of damage of the
surface in terms of a "flaw distribution function".   In principle, this function can be used to predict the strength of the
glass under any specified system of loading.   Only the most
elementary account of flaw statistics will be given here.

Consider a cylindrical glass rod, one individual from a
large population of similar rods, which is subjected to a uniform axial tensile stress.   Let this stress be gradually increased from zero.   Below a certain stress σ_u, the rod will
not fracture, nor will any rod from the population.   There are
no flaws in any of the rods which are severe enough to act as
a fracture origin at a stress less than or equal to σ_u.   At
higher values of stress there will be an increasing number of
flaws to be found in the rods which are capable of acting as
fracture origins.   The symbol $n(\sigma)$ will be used to denote the
mean number of flaws per unit area which are capable of acting
as fracture origins at stresses $\leqslant \sigma$.   This is the flaw distribution function.   It has the value zero at stresses less than
σ_u.

Now consider a very small element of the rod surface of
area δA such that the product $n(\sigma).\delta A$ is less than 1.   This
does not mean that the area contains a fraction of a flaw.   A
more reasonable interpretation is that the product is a measure
of the probability, p, of the element containing a flaw capable
of acting as a fracture origin at a stress less than σ.   Any
element that does contain a flaw will, of course, fracture at
some stress $\leqslant \sigma$.   On the other hand there is a probability
q = 1 - p that the element will withstand the stress without
fracture.

The finite surface area, $A = 2\pi rl$, of a rod specimen is made up from a large number, m, of such elements ($m = A/\delta A$). The important question in practice is "What is the probability of finding a flaw which can act as a fracture origin within a finite area of this size?" As it happens, it is far easier to answer the complementary question "What is the probability of not finding such a flaw?" or "What is the probability of the area A surviving when subjected to a specified stress, σ?"

This survival probability, Q, is given by evaluating the probability of all m elements surviving simultaneously. Now the probability of occurrence of m simultaneous events, each event having the same probability, q, is q^m. Therefore

$$Q = q^m = (1-p)^m$$
$$= (1-n(\sigma).\delta A)^m$$
$$= (1-n(\sigma).\frac{A}{m})^m$$

It is easy to show that when m is very large, $Q = \exp(-n(\sigma).A)$. Thus the probability that fracture of the rod will have occurred by the time the stress has been increased to σ, is

$$P = 1-Q = 1 - \exp(-n(\sigma).A) . \tag{64}$$

If we keep σ and hence $n(\sigma)$ constant, and consider what is the effect of increasing A, the area subjected to stress, we find that the fracture probability P increases. When the product $n(\sigma)A = 1$, $P = 0.63$ and when $n(\sigma).A = 4$, $P = 0.99$.

For the relatively simple case of glass rods or fibres subjected to axial tensile stress, it is easy to determine $n(\sigma)$ experimentally. The fracture stress results for a large sample of specimens may be plotted as a cumulative frequency distribution, i.e. a graph showing how the fraction of samples which have broken increases with the applied stress. This graph will have the form shown in Fig. 64. Now F is a direct experimental measure of the fracture probability, P. Thus if we obtain Q from the experimental results using the relationship:

$$Q = 1-P = 1-F,$$

then $n(\sigma)$ is easily obtained from the equation
$$n(\sigma) = - \ln Q/A.$$

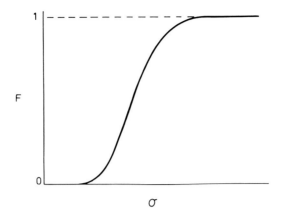

Fig. 64.   Cumulative distribution of strength measurements.

In most strength measurements, the stress varies over the surface of the glass.   It is then more difficult to derive the flaw distribution function from the results.   Figure 65 shows

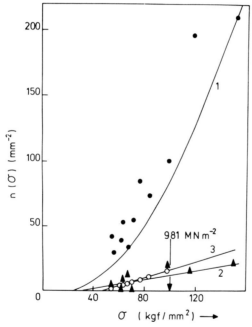

Fig. 65.   Flaw distribution function for three types of glass surface.   (1) Ground and polished plate.   (2) Float glass, tin-contact surface.   (3) Float glass, upper surface.

the variation of n(σ) with σ for two float glass surfaces

(Hamilton and Rawson, 1972).   These flaw distribution functions
were calculated from Hertz fracture strength measurements and
for that reason they do not show the presence of the most severe
flaws.

An interesting practical important application of flaw statistics has recently been described by Olshansky and Maurer (1976).
Glass fibres up to several km in length are coming into use in
telecommunication systems.   It is necessary to ensure (on the
basis of strength measurements carried out in the laboratory on
relatively short lengths) that multi-kilometre lengths can withstand short-term tensile stresses up to 700 Nmm^{-2} and lower
stresses for periods up to 20 years.

3.   Time Effects

a.   *Static Fatigue*.   Imagine a stress applied instantaneously
to a glass specimen.   If the stress is high enough, the glass
will fracture immediately the stress has been applied.   If, on
the other hand, the stress is very low, the glass will not fracture, no matter how long the stress is applied.   At intermediate
stresses it is found that the glass does not fracture immediately,
but it supports the stress for only a limited time;   the higher
the stress, the shorter is the time to fracture.   This effect
is known as static fatigue.   Its magnitude depends on a number
of factors, in particular the ambient conditions and the glass
composition.

A very detailed study of the effect has been carried out by
Charles (1958a, 1958b), who also developed a theory to explain
his observations.   He determined the effect of the applied stress
on the time to fracture of centreless-ground rods of soda-lime-
silica glasses loaded in three-point bending.

The atmosphere surrounding the rods was air saturated with
water vapour, and the magnitude of the static fatigue effect was
determined at various temperatures in the range -170°C to 242°C.
Each experiment gives a result of the type shown in Fig. 66 - a
distribution of fracture times.   For this particular experiment,
involving a batch of about 60 rods, each rod was subjected to
the same forces, so as to produce a maximum stress at the centre
of the rod of 52 MNm^{-2}.   The temperature of the experiment was
24°C.   It may be seen from the figure that the most probable

127

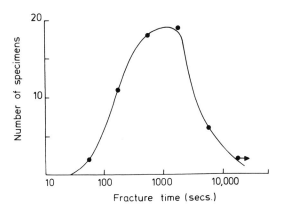

Fig. 66.   Distribution of fracture times.

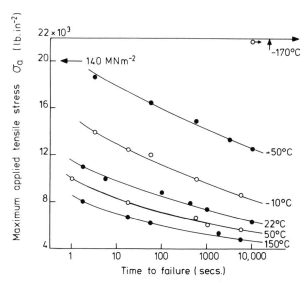

Fig. 67.   Relation between fracture stress and time to failure
at various temperatures.

fracture time was approximately 1000 s.   Figure 67 combines the
results from many such experiments, and shows the relationship
between the most probable fracture time and the maximum applied
stress at various temperatures.   It is particularly important
to note that there is no static fatigue effect in the experiments carried out at -170°C.

The explanation of static fatigue given by Charles is that
chemical reaction occurs between water vapour in the atmosphere

and the glass. The rate of the reaction is much increased by the high stress in the glass at the tips of surface flaws. As a result of the reaction, the flaws become deeper and eventually the crack depth, c, becomes equal to the critical depth, c*, which satisfies the Griffith criterion for fracture at the given applied stress (Equation (49)).

Charles is careful to point out that the stress-accelerated reaction does not necessarily result in the flaw becoming more severe, i.e. in the stress at the flaw tip increasing. The Inglis equation shows that this stress depends on the ratio of crack depth to crack width, c/b. If the theoretical fracture stress is assumed to be 1.4×10^4 MNm^{-2}, a value of c/b = 40 would be required to produce this stress at the flaw tip if the applied stress were 180 MNm^{-2}. If a lower stress of 36 MNm^{-2} were applied to a glass containing a flaw of c/b = 40, for fracture to occur it would be necessary that the flaw should change its shape, as a result of the stress-accelerated reaction, until c/b reached a value of 200, i.e. it should grow more rapidly in the "c" than in the "b" direction. If this condition were not satisfied, the glass would become stronger as a result of the stress-accelerated reaction.

At the time Charles developed his theory, no information was available on how the velocity of crack growth v, varies with the applied stress and the temperature of the experiment. Charles assumed the following equation for V

$$v = dc/dt = A(\sigma_y{}^n + B).\exp(- E/RT) , \qquad (65)$$

where A, B and n are constants, σ_y is the stress at the flaw tip, T is the temperature in °K and E is the activation energy for the temperature-dependent reaction of water vapour with the glass.

This expression was used to obtain an equation for the time, t, required for a flaw of initial length c to grow to the critical length c* for fracture at the applied stress, σ. At a given temperature, the equation is

$$\log t = n.\log(1/\sigma) + b \qquad (66)$$

Figure 68 shows that the results fit this equation very well, at least in the temperature range -10° to 150°C. A value

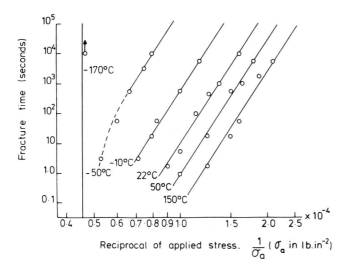

Fig. 68. Data of Fig. 67 replotted to test Equation 66.

of n = 16 was determined from the slope of the straight lines,
showing that the velocity v increases very rapidly with increasing stress.   It was also possible to obtain from the results a
value for the activation energy, E.   The value obtained, 18.8
Kcal mole^{-1}, was very close to that obtained from experiments
on the effect of temperature on the rate of attack of the same
glass by water vapour.   Charles suggested that the reaction
rate was controlled by diffusion of sodium ions within the glass
structure.

A greal deal of work has been done in recent years pioneered
by Wiederhorn and his colleagues (Wiederhorn and Bolz, 1970;
Wiederhorn et al.,1974a), in which the growth of microscopic
cracks in glass has been measured under a wide range of conditions.

Figure 69 shows results obtained on a soda-lime-silica glass,
the test specimens being in the form of microscopic slides.   The
effect of increasing relative humidity in increasing the growth
velocity is obvious.   At low values of the stress intensity
factor, the relationship between v and K_I is of the form

$$v = A.K_I{}^n ,$$

a result slightly different from the form assumed by Charles.
This is followed by a range of K_I values, (II) in which v is

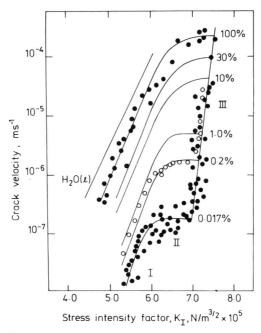

Fig. 69. Effect of relative humidity on the relationship between crack velocity and stress intensity factor for a soda-lime-silica glass.

independent of K_I. Wiederhorn suggests that, in this region, v is controlled not by the rate of chemical reaction at the crack tip, but by the rate of diffusion of water vapour along the crack At high values of K_I, v become independent of the moisture content of the surroundings (range III). Measurements carried out in vacuum (Wiederhorn et al., 1974a) suggest that in region III, the fracture process may involve a mechanism which depends upon the elastic properties of the glass and hence upon the glass structure.

b. *Proof testing*. The detailed measurements of crack growth in glasses and their analysis in terms of fracture mechanics has led to a potentially valuable method for predicting the minimum service life of a glass component under stress. This is known as "proof testing" (Wiederhorn and Evans, 1974). The application of the technique to the design of space shuttle windows is described by Wiederhorn et al., (1974b). The lives of the shuttle crew depend upon the windows remaining intact, and it is

for such critical applications that the technique is likely to
be most used.

In carrying out a proof test, the component is subjected to
a stress which is relatively high, but not high enough to cause
fracture.   The proof test should be of brief duration and pre-
ferably be carried out in vacuo, so that there is negligible
crack growth during the proof test itself.   If the glass sur-
vives the proof test, in which the applied stress is σ_p, this
sets an upper limit on the size of the most severe flaw present.
The minimum time to fracture under a long-term service stress,
σ_s, can then be calculated by integrating the equation for the
crack velocity from the maximum initial length, determined by
the proof test, to the critical length at which the crack is
energetically unstable at the stress σ_f.

The analysis is made relatively simple by the fact that for
most of the time during which crack growth is occurring, the
equation for region I can be applied, i.e. $v = A.K_I^n$.   The value
of n is always large, between 15 and 50 for oxide glasses.

During service, the most severe crack grows from an initial
length c to its final critical length c*, and K_I increases from
its initial value, K_{Ii}, to the critical value, K_{IC} .

From

$$K_I^2 = Y^2 . \sigma_s^2 c \text{ and } v = dc/dt$$

we obtain

$$v = dc/dt = (\sigma_s^2 . Y^2)^{-1} . 2K_I . dK_I / dt . \qquad (67)$$

The time for crack growth, t, is obtained by re-arranging this
equation and integrating,

$$t = 2(\sigma_s^2 . Y^2)^{-1} \int_{K_{Ii}}^{K_{IC}} (K_I/v) dK_I . \qquad (68)$$

Substituting for v and evaluating the integral,

$$t = 2(K_{Ii}^{2-n} - K_{IC}^{2-n})/(A\sigma_s^2 . Y^2 (n-2)) \qquad (69)$$

Since n is large, K_{IC}^{2-n} is very small compared with K_{Ii}^{2-n} .
Hence

$$t = 2K_{Ii}^{2-n}/(A\sigma_s^2 . Y^2 (n-2)) \qquad (70)$$

Thus the evaluation of the minimum service life depends upon
evaluating, from the proof test, the value of K_{Ii} corresponding
to the most severe flaw originally present in the glass. K_{Ii}

cannot be greater than the value calculated from $K_{Ii} = Y\sigma_p c^{\frac{1}{2}}$ where c is the initial length of the most severe flaw. If σ_f is the stress at which fracture would occur if no crack growth occurred

$$K_{Ii} < K_{Ic} = Y\sigma_f . c^{\frac{1}{2}} .$$

Also

$$K_{Ii} < K_{Ic} = Y\sigma_s . c*^{\frac{1}{2}}$$

where c* is the final length of the crack at which fracture would occur under the service stress, σ_s.
Hence

$$K_{Ii} < (\sigma_s/\sigma_p) K_{Ic}$$

Combining this with Equation (70), the service life will be greater than t_{min} given by

$$t_{min} = 2(\sigma_s/\sigma_p)^{2-n}.K_{Ic}^{2-n}/(A\sigma_s^2.Y^2(n-2)) .$$

Using this equation, a diagram can be constructed giving the re-lationship between the minimum service life and σ_s for various values of σ_p/σ_s. Figure 70 shows one such proof stress diagram for soda-lime-silica glass immersed in water. The use of the diagram may be illustrated by taking an example. Suppose it is required that the minimum service life ahould be 10^6s under a long-term service stress of 50 MNm^{-2}. The diagram shows that, to guarantee this the glass should be subjected to a proof test at a stress of 200 MNm^{-2}.

c. *Dynamic fatigue.* When stress is applied to glass at a con-stant rate, it is found that the fracture stress is higher, the higher the rate of loading. This effect, called dynamic fat-igue, was also investigated by Charles (1958c), some of whose results are shown in Fig. 71.

The effect can easily be understood in terms of the ideas put forward in the previous section. When the rate of loading is low, there is more time for reaction with water vapour at the flaw tip to occur. Hence there is more crack growth during loading. At low temperatures, when the concentration of water vapour in the atmosphere is very low, the effect is not detect-able; even at room temperature it is not very large. A thou-sand-fold increase in the rate of loading increases the frac-

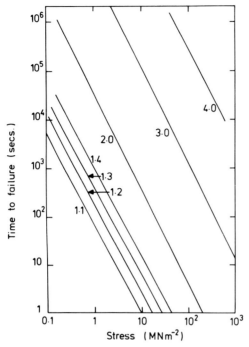

Fig. 70.   Proof test diagram for soda-lime-silica glass.

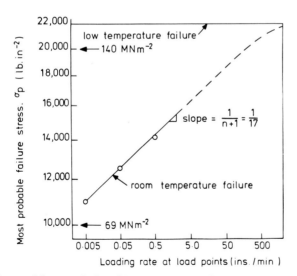

Fig. 71.   Effect of loading rate on fracture stress.

ture stress by only about 50%. However, when measuring glass
strength, one should standardise on a rate of loading, which
should be stated when presenting results.

Charles showed that, when the applied stress increases at
a constant rate β, the fracture stress is related to β by

$$\sigma_f = K\beta^{(1/(n+1))} \tag{72}$$

K is a constant and n is the exponent in Equation (65). The
value of n can be determined from the slope of the straight line
which is obtained when $\log\sigma_f$ is plotted against $\log\beta$ (Fig. 71).
Charles obtained a value of n = 16, a value in agreement with
that obtained from his static fatigue experiments.

In more recent work, Ritter and Sherbourne (1971) show how
Charles' theory and results are supported by measurements of
Wiederhorn and Bolz (1970), who have investigated the effect of
stress on the rate of growth of macroscopic cracks in glass in
the presence of water vapour.

G. Increasing the strength of glass

As pointed out earlier, high strength glass specimens may
be produced by removing the surface flaws. However, this is
of little practical value, since new flaws can be produced so
easily. Practical methods of increasing the strength depend
upon preventing existing flaws from growing. Since crack growth
can occur only when there is a tensile stress at the flaw tip
acting at right angles to the plane of the crack, strengthening
can be achieved by prestressing the glass so that in the un-
loaded state it contains a surface compressive stress acting in
directions parallel to the glass surface, and over a distance
from the surface which is significantly greater than the length
of the deepest surface flaw. Before the tip of the flaw can
be subjected to tensile stress, it is then necessary to apply
loads which are high enough to neutralise the pre-existing surface compressive stress. In this way the strength of glass
components can be increased by a factor of about five.

The strengthening compressive stresses may be produced
either by a thermal or chemical treatment. Thermal strengthening or "toughening" has been widely used industrially since the
late 1920's, but the fact that glass could be strengthened by

heat treatment has been known for at least three hundred years.
Approximately spherical droplets of glass a few millimetres in
diameter having a long thin tail can be made by carefully pour-
ing a thin stream of molten glass into water.   These are the
so-called "Prince Rupert's drops".   The bead portion is extremely strong, it is very difficult to cause fracture even by
striking the bead with a hammer.   However the tail is easily
snapped.   When this is done the whole droplet fractures explo-
sively into a fine powder.   This is similar to the rather less
explosive fracture which occurs when a toughened car windscreen
breaks.   The glass fractures into a large number of approximately equiaxed fragments a few millimetres in size.

Toughened windscreens and oven doors are made from commercial flat glass by suspending the glass, usually vertically, so
that it does not deform under its own weight.   The sheet is then
heated in a furnace to a temperature between 750 and 800°C.
Each surface of the glass is then rapidly cooled by air jets.
To obtain a more uniform stress distribution, the points of impact of the jets are moved over the glass surface during the
cooling process.   The way in which this treatment leads to surface compressive stresses can be understood from the account of
the annealing process given in the previous chapter.

The stress distribution through the thickness of the glass
is approximately parabolic, and the thickness of the surface
compression layer is considerably greater than the depth of the
most severe flaw.   There are, of course, balancing tensile
stresses in the interior of the sheet.   As a result of the high
toughening stresses, there is a considerable amount of strain
energy stored in the glass.   When fracture occurs, this energy
is rapidly liberated and appears largely in the form of the surface energy of the fragments produced.   Hence the surface area
produced is large, accounting for the small size of the fragments.
The lacerating effect of these fragments on the body of the car
driver is far less than that of the large dagger-like fragments
formed when untreated glass breaks.   It is on this, rather than
on the increased strength, that the safety feature of the toughened windscreen depends (Kay 1973, Lister 1961).   Although common in the United Kingdom, the thermally toughened windscreen is
not used in the United States where the laminated type of wind-

screen is required by law. This consists of two sheets of
glass with an interlayer of a polymer (usually polyvinyl butyral).
Except under conditions of very severe impact, the large glass
fragments are held together by the much more elastic polymer
layer. In recent years practice in the two countries has shown
some signs of coming together by the introduction of a type of
windscreen in which layers of glass, strengthened either thermally or chemically, are formed into a laminated screen. The
sheet on the driver's side of the screen is more highly pre-
stressed and fractures into relatively small pieces. At the
same time, use is made of the energy absorbing property conferred
by the polymer interlayer (Blizard and Howitt, 1969/70).

Thermal toughening is also used for strengthening pressed
glass tumblers and glass power-line insulators. However, articles of complicated shape are more difficult to toughen thermally
than glass sheets. It is also difficult to use the thermal process if the glass is only one or two millimetres thick, since the
magnitude of the stresses is proportional to the temperature
gradient produced by rapid cooling. It is difficult to produce
large temperature gradients if the glass is thin.

The chemical techniques for strengthening glass are not subject to the same limitations, although they have the disadvantage
of being rather slow compared with thermal toughening. The
chemical methods have been exploited commercially only during
the last ten years or so, but well before that the glass industry was using a chemical strengthening process, not always fully
realising the fact. Before the introduction of modern continuous annealing lehrs, heated by electricity, either town gas or
oil was used. The combustion gases contain SO_2 which reacts
with the alkali in the surface layers of the glass to form
Na_2SO_4. The glassware merging from the lehr was covered with
a white bloom, which had to be removed by washing. This reaction resulted in the formation of a surface layer on the glass
of reduced alkali content and hence of lower expansion coefficient. Thus a layer of compressive stress was set up in the
glass surface, the thickness of which is of the same order as
that of the chemically modified surface layer.

The mechanism of the reaction and the increase in strength
which it produces has been studied by a number of workers, most

recently by Mochel et al. (1966) . An indication of the extent
by which this sulphating treatment may increase the strength can
be obtained from some of the results which they reported. Thus
the mean value of the modulus of rupture of a batch of glass rods
which had been given the sulphating treatment was 19,400 psi
compared with 16,900 psi for untreated rods.

In a newer and more widely used method of chemically strengthening, the glass is immersed in a fused salt mixture containing
potassium ions at a temperature well below the annealing point
of the glass. Base exchange occurs at the glass surface, sodium
ions diffusing from the glass into the fused salt being replaced
in the glass by an equal number of larger potassium ions. This
results in a compressive stress being set up in the glass surface.
The effect is due not to a thermal expansion difference, but to
the difference in size between the alkali ions. The larger
potassium ions move into sites in the rigid silicate network
previously occupied by sodium ions. This tends to expand the
glass structure, a tendency which is resisted by the chemically
unchanged interior layers. The greatest increases in strength
are obtained using an alumino-silicate glass composition, in
which alkali ion diffusion occurs more rapidly than in a conventional soda-lime-silica glass. Strengths of approximately
500 MNm^{-2} may be obtained by this method, which has been used
most extensively in treating lenses for safety goggles and also,
in the United States, for treating the component glass sheets
of laminated windscreens.

Good accounts of the method have been given by Zijlstra
and Burggraaf (1968) and by Garfinkel and King (1969, 1970).

The distribution of stresses through the glass thickness is
quite different from that obtained by thermal toughening (Fig.
72). The surface compression layer is much thinner and in the
interior of the glass the tensile stresses are much lower.
Consequently, the strain energy stored in the glass is less
than when thermal toughening is used and larger fragments are
formed when chemically strengthened glass is broken.

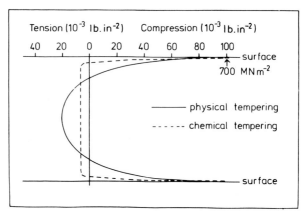

Fig. 72. Distribution of stresses in thermally toughened and chemically strengthened glass.

CHAPTER V

THE STRENGTH OF GLASSWARE

The account given in the previous chapter of the strength
of glass as a material has shown that it depends on many factors.
When one considers the strength of a particular article of glassware (a container, a window, or a lens for safety goggles) additional factors have to be considered.   In particular one needs
rather detailed information on the distribution over the glass
surface of the stresses to which the article is subjected during
its life.   This information should include the effects on the
surface stress distribution of the variations in shape and thick-
mess of the article which are likely to be encountered in its
manufacture.   When considering the implications of this information, it is necessary to remember that the strength of the article will probably decrease during use as a result of surface
damage.

Although considerable progress is now being made in the
application of numerical methods to the calculation of stresses
in glass articles, it is still necessary to rely to a considerable extent on experience, and on the use of simulated service
tests carried out in the laboratory.   Some such testing is essential since, as has been pointed out in the previous chapter,
the initial strength of a glass surface depends on the surface
damage which the article has received during manufacture.   The
extent of damage is not uniform and, for articles of complicated
shape like glass containers, the most severe damage may move
from one region to another during a production run.   Simulated
service testing is most widely practised in the glass container
industry, where it is a normal and essential part of quality
control.   The tests may include simulated abrasion of the containers using a machine in which they are subjected to damage
conditions similar to those encountered on a high-speed filling
line.   By measuring the strength of the containers before and
after treatment on such a line-simulator, a good indication is
obtained of the susceptibility of a particular design to damage
and of the probable strength of the container when it reaches
the user.

In spite of the considerable effort devoted to design and

testing, fractures of glass articles do occur, and sometimes in-
jury to the user results.   It is important in such cases to be
able to assess whether the fracture was the result of a manufacturing defect or whether it was due to the user maltreating the
glass in some way.   This type of question can often be answered
by the techniques of fracture analysis described at the end of
this chapter.

A.   Determination of Service Stresses

1. Glass Containers

Containers used for beer and carbonated drinks, ranging in
quality from "Coca-cola" to champagne, have to withstand internal pressure.   For beer and soft drinks the highest pressures
are usually encountered in the bottling plant.   According to
Moody (1963), the highest internal pressure to which a beer
bottle is subjected is 80 lb in^{-2} (0.56 MNm^{-2}) during pasteur-
ization at 66°C.   This falls on cooling to 20 lb in^{-2}.   During
the filling of bottles with carbonated drinks, the pressure may
rise to 60 lb in^{-2} and similar pressures may be encountered
during storage in warm conditions.   Moody states that the maximum storage temperature likely to be encountered in the U.K. is
about 38°C, but temperatures as high as 49°C may be encountered
in hotter countries.

The pressure encountered during storage depends mainly on
the storage temperature and the degree of carbonation; i.e.
the number of volumes of CO_2 dissolved in one volume of liquid
at N.T.P.   A secondary but important factor is the volume of
free space left in the container, termed the vacuity.   Moody
states that the vacuity should never be less than 2.5% of the
container volume, or pressures may become excessively high.
He gives a table for the pressures corresponding to various temperatures and various degrees of carbonation.   Measurements of
internal pressures have been made by Hopper (1968).   They are
shown graphically in Fig. 73.   Good agreement exists between
measured pressures and values calculated from the gas laws and
data for the solubility of CO_2 in water (Bowles, 1976).

The stresses in the wall of the container due to the internal pressure vary considerably from one part of the container
to another.   It is relatively easy to make a good estimate of

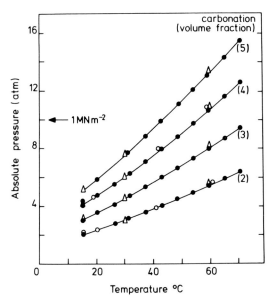

Fig. 73. Effect of temperature and degree of carbonation on the pressure in a container.

the stresses in the central section of a container of cylindrical cross section. It is in this section that fracture due to internal pressure normally occurs. Treating it as a thin walled cylinder in which one may ignore variations in stress through the thickness of the wall, one can easily show that the principal stresses are: a circumferential tensile stress σ_θ, and an axial tensile stress σ_z given by

$$\sigma_\theta = pR/t \qquad\qquad (74)$$

$$\sigma_z = pR/2t \qquad\qquad (75)$$

where R is the radius, t the wall thickness and p the internal pressure. Thus the maximum tensile stress in a container of 30 mm radius and 2 mm wall thickness is 10.5 MNm^{-2} when the internal pressure is 100 lb in^{-2} (0.7 MNm^{-2}). Normally the fracture stress of the glass is expected to be 50-100 MNm^{-2}, indicating a fairly high factor of safety. The reader is referred to the revised edition of Moody's book (1977) for more detailed information on the strength and design of containers.

Until recently, only experimental methods were available for determining the overall stress distribution in an article as

142

complicated as a glass container. The same was true for the determination of the even more complicated distribution in a television tube, which has to withstand for very long periods the stresses produced by evacuation of the tube. An excellent account of the application of experimental methods to the determination of stress distribution in containers has been given by Teague and Blau (1965). They used strain gauges cemented to the glass surface, photoelastic measurements carried out on plastic models, and the brittle lacquer technique. The latter is a useful qualitative method for drawing attention to regions of high tensile stress. The article is sprayed with a polymer solution which hardens to a brittle coating. When the article is stressed, cracks develop in the coating in regions where the stresses are high, the cracks running at right angles to the maximum tensile stress.

Experimental investigations of this type are very time-consuming, especially if one wishes to make a study of the effects of design changes. Fortunately computer programs are now available, usually based on the finite element method of stress analysis, by which pressure stresses in glass containers and TV tubes can be calculated. Using these programs, the effects of design changes can readily be investigated.

So far as the writer is aware, it is not yet possible to compute the stress distribution produced in a hollow glass article by impact or by thermal shock. One or two experimental studies of the stresses produced by impact have been made. They have used strain gauges cemented at various positions to the glass surface, the gauges being connected to high speed recording equipment (Budd and Cornelius, 1976). Apparently no experimental measurements of thermal shock stresses have been made. There seems to be no reason however why it should not be possible to compute the stresses due to thermal shock using information, which would have to be obtained experimentally, on heat transfer conditions at various points over the glass surface.

2. Glass Windows

The normal type of loading for which glass windows have to be designed is that produced by wind pressure. It is obviously

essential that careful consideration be given to the design of
large windows in the upper floors of high buildings. At first
sight one might imagine that the problem of calculating the
stresses due to a uniform wind pressure on a rectangular window
would be easy compared with that of calculating the pressure
stresses in a glass container. Unfortunately this is not so.
The problem of calculating stresses in rectangular plates subjected to bending forces is one of the most complicated in the
area of the theory of plates and shells (Timoshenko and Wainow-
sky-Krieger, 1959). Much of the difficulty arises from the fact
that the stresses depend very much on how the plate is supported
or gripped at its edges. As for containers, current practice
is based very much on field experience and on carefully con-
ducted laboratory experiments. However, this again is a problem for which numerical methods of stress calculation are beginning to make a valuable contribution (Tsai and Stewart, 1976).

Since the specification of the thickness of glass to be
used in a window is a matter for the builder or architect rather
than for the glass manufacturer, it has been necessary for manufacturers of flat glass, or associations of manufacturers, to
publish recommended codes of practice. These vary somewhat
from one country to another. A good indication of the nature
of these codes can be obtained from the Glazing Manual published
by the U.S. Flat Glass Manufacturing Association (1974). This
includes a wind map and tables giving the highest wind velocities and wind pressures encountered at various altitudes throughout the U.S.A. Figures 74 and 75, based on this information,
give the wind pressure as a function of surface wind velocity
and height above ground level. From the pressure which a window of specified area must withstand, the thickness of glass to
be used may be read from a table. Figure 76 shows a number of
design curves based on this table.

The dynamic pressures and stresses developed in windows
by sonic booms from aircraft have been studied by Welch and
Hudson (1976). According to the U.S. Flat Glass Manufacturing
Association Manual, air pressures greater than 20 lb ft^{-2} are
unlikely to result from supersonic aircraft operating within
F.A.A. regulations.

Considerably more complex stress distributions have to be

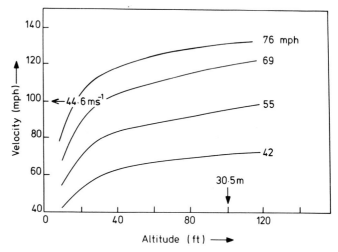

Fig. 74. Effect of altitude on wind velocity for various surface wind velocities.

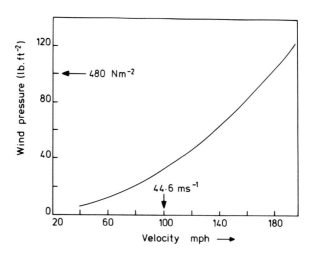

Fig. 75. Relation between wind pressure and wind velocity.

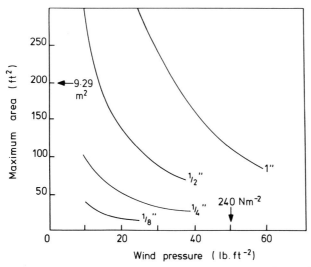

Fig. 76. Maximum recommended window area as a function of
wind pressure for various glass thicknesses.

considered when designing windscreens for cars and aircraft.
In the design of a car windscreen, the aspect of importance is
not so much the strength of the windscreen as its effect on the
human head when striking it at a high velocity.   The aims in
designing the windscreen are to minimize the peak impact forces
and the degree of laceration which occurs when the screen breaks.
Only experimental work can provide this information.   The equipment used is quite complicated (Blizard and Howitt, 1969, 1970).
A human dummy mounted on a sledge designed like a car seat is
propelled towards a frame holding the screen under test.   The
head of the dummy is designed to have a resilience similar to
that of human flesh and bone and is covered with soft leather,
the cutting of which by the fragments of the screen gives a measure of the degree of laceration to be expected.   Accelero-
meters mounted in the head of the dummy measure the impact
forces.   Experiments such as this have led to the introduction
of new types of windscreen using pre-stressed glass in combination with a polyvinyl butyral interlayer (Blizard and Howitt,
loc. cit.; Kay, 1973).   These considerably reduce impact in-
juries compared with the designs previously available.

For aircraft windows, a major hazard is collision with flying birds, especially during take-off and landing. Here the strength of the window is of primary importance. Again, simulated service tests provide the only method of obtaining the necessary information. They involve the rather unpleasant procedure of firing bird carcasses at the screen under test, using an air cannon.

An excellent account of a theoretical and experimental study of the stresses produced by impact of small high velocity particles on lenses used in safety goggles has been given by Goldsmith and Taylor (1976).

B. The Simulated Service Testing of Glass Containers

The simulated service testing of glass containers is further developed than for any other type of glassware. There are national standards describing the test procedures, and equipment for carrying out the tests is commercially available. The standard test procedures describe only the methods to be followed and some general features of the apparatus to be used. The performance of containers when subjected to the tests is not laid down. This is a matter for discussion between the manufacturer and the customer.

The two tests which are most widely used are a thermal shock test and a bursting pressure test. A thermal shock test is described in A.S.T.M. specification C 149-50. The apparatus consists of two large tanks, one containing hot water, the temperature of which can be controlled, and the other containing cold water. The containers are filled with hot water from the first tank and are placed in a metal frame basket which is immersed in the water. The basket is designed in such a way that the bottles are kept separate so that water has free access to the entire outer surface of each container. The containers are left for five minutes in the hot tank to reach a steady temperature. They are then quickly transferred to the cold tank, after which they are inspected and a note made of the number of containers which have fractured. The method of test may involve a single thermal shock experiment or a series of tests in which the temperature of water in the hot tank is progressively increased. The latter type of test is more informative, but a single shock test may be acceptable for determining whether or not a particular batch of containers is acceptable.

The bursting pressure test is described in A.S.T.M. specification C 47-62. Each container in the sample is filled with water and is mounted by its neck in a machine which increases the pressure on the water in a controlled way. The pressure may be increased up to a specified value and a note made of the number of containers surviving, or it may be increased to the bursting pressure of each container. The specification requires that the pressure should be increased at a rate of at least 10,000 lb in^{-2} min^{-1}. In the progressive test the pressure is increased in a series of steps. After each pressure increase, the pressure is held constant for a predetermined period.

Other simulated service tests which are widely used in the container industry, although not defined by national standards (at least, not in the U.K. or U.S.A.) are an impact test and a crushing test. In the impact test the container is firmly supported vertically and the impact is supplied by a ball-ended hammer which swings on a pivoted arm in a vertical plane to strike the container on its side. The impact is measured by the angle through which the arm is raised before it is released. The vertical crushing test is used to measure the ability of the container to withstand the force exerted by certain designs of capping machine.

C. Fracture Analysis

Careful examination of fracture fragments can often be used to determine the cause of fracture and to identify the fracture origin or origins. There are at least two situations where it is valuable to be able to do this. If an abnormally large proportion of glassware breaks on the production line, it is obviously important to determine the reason. Similarly, if during any of the strength tests carried out as part of the quality control procedure, there is a marked change from the pattern of fracture normally observed in that test, this may indicate that a region of weakness exists in a region of the article where that weakness did not exist previously. Another very important area where fracture analysis is extremely valuable is in dealing with customer complaints. The customer may be a company

filling containers with food or drink which runs into a period
of unusually high failures on the filling line, or it may be an
individual who has been injured by the fracture of a glass article under conditions which that individual considers the glass
article should not have fractured. When this happens, claims
for damages may be involved. It may be difficult to come to
any very definite conclusion about the cause of fractures because very often in such situations all the fracture fragments
are not recovered.

There are two aspects to fracture analysis. One involves
an examination of the general form of the fracture pattern, and
the other a careful and detailed examination of the fracture surfaces. In order to obtain a clear picture of the fracture pattern, it is often necessary to stick the fragments together in
the form of the original article. This is not too difficult
if the fragments are relatively large, and involves the use of
adhesive tape and modelling clay.

There are a number of general points to be made about the
fracture pattern:

(i)   Even though the pattern may be quite complicated and
        shows a considerable amount of forking, there is very
        often only one fracture origin. Two notable exceptions
        to this statement are fractures produced by very severe
        thermal shock, and fractures produced by impact on the
        cylindrical side wall of a glass bottle. In both
        instances multiple origins may be found.

(ii)   The fracture origin is nearly always located at the
        glass surface, although if there is a small piece of
        undissolved refractory inside the glass, this may
        act as an origin.

(iii)   The fracture propagates in a direction at right angles
        to the maximum tensile stress at the tip of the fracture.

(iv)   Fractures which propagate at low stress levels move
        slowly and do not fork, whereas those propagating under
        high stresses move very rapidly and show considerable
        forking. Figure 77 is a high speed photograph of a
        fracture produced by a pellet fired at the centre of
        a sheet of toughened glass. Using such photographs
        it has been shown that there is a maximum fracture

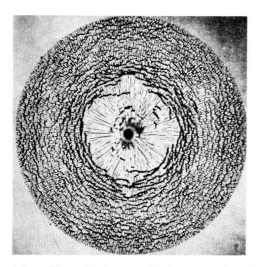

Fig. 77. High speed photograph of glass fracture.

velocity which depends on the elastic properties and
the density of the glass. This velocity for most
oxide glasses lies between 750 and 2155 m s^{-1}
(Schardin, 1959).

We may now consider fracture patterns which illustrate some
of these general points. The effect of the magnitude of the
stress on the fracture pattern is easily shown by breaking glass
rods by bending. If the rod is heavily abraded before it is
broken, a simple transverse break is observed originating from
the rod surface and on the tension side of the neutral axis
(Fig. 78). In fracture of an unabraded rod, the fracture usually forks and the rod breaks into at least three pieces.

In the previous chapter a description was given of the
Hertzian fracture produced when a small steel ball is pressed
against a flat surface. The fracture origin is on the glass
surface at a point just outside the circular area of contact
between the ball and the glass. The region of the fracture
nearest to the surface in contact with the ball is cylindrical
and is formed by the fracture running around the circle of contact and down into the glass in a direction at right angles to
the maximum tensile stress, which is predominantly radial.
The conical region formed further from the surface can also be
explained in terms of the stress distribution produced by this
type of loading (Frank and Lawn, 1967).

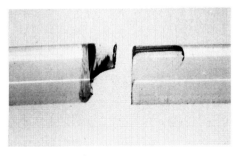
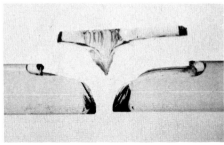

a) abraded   b) unabraded

Fig. 78.   Fracture of glass rods

A small spherical pellet or a stone striking a glass window at a high velocity often produces a Hertz fracture which may
propagate through the sheet so that a conical fragment of glass
is punched out.   Impact on a glass sheet by a larger and more
slowly moving object produces a quite different type of fracture.
Then the glass will bend under the impact, and tensile bending
stresses are set up on the surface opposite the region of impact
(Glathart and Preston, 1968).   An approximately radial pattern
of cracks will then form, since the largest tensile stresses are
circumferential.   The crack pattern is similar to that seen in
glass discs which have been fractured in bending (Fig. 49).

Fracture of a glass container by impact is more difficult
to interpret.   In impact testing, where a heavy round-nosed
hammer strikes the wall of a rigidly supported container, one
may see radial cracks starting from a point on the internal sur-
fact close to the impact point.   In addition there may be at
least two other origins some 90° to 120° around the wall on each
side of the impact point.   These are due to the bending stresses
produced by the deformation of the cross-section of the container
(Mould, 1952, Budd and Cornelius, 1976).   When dealing with
such a complex pattern it is difficult to identify what is the
normal pattern for a particular design of container.   The pattern may not be the same when the container is full as when it
is empty.   A simple method of preserving the fracture pattern

so that it can easily be inspected is to wrap most of the surface with adhesive tape before the test, leaving bare the area
to be impacted (Fig. 79C).

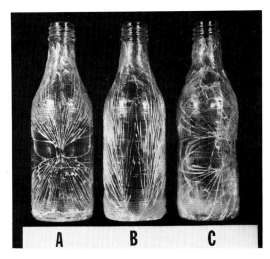

Fig. 79. Fracture of containers which have been broken by
internal pressure (A and B) and by impact (C).

The fracture produced by subjecting a cylindrical container to a high internal pressure usually takes a very characteristic form (Fig. 79A). As shown earlier in this chapter (Equations (74) and (75)), the tangential stress is the highest tensile stress and the initial crack runs axially from the origin.
As the fracture accelerates it begins to fork and one observes
a fan-shaped array of cracks above and below the fracture origin.
These eventually run around the neck and the base of the bottle.
Note however that this pattern is not always observed (Fig. 79B).

The stresses normally encountered when a glass container
fails due to thermal shock are usually not high enough to cause
forking. When a cylindrical container containing hot water is
plunged into cold water, tensile stresses develop on the outer
surface. The fracture usually starts at a point near the base
and first runs around the base in a direction parallel to the
base (Fig. 80). Just before the fractures join, one crack de-
viates and runs in an axial direction up the container wall.

The design of glassware is influenced by what is known
about the normal mode of failure under service conditions. An

152

Fig. 80. Fracture of a container which has been broken by thermal shock.

attempt is made to design the article in such a way as to protect from damage the region in which fracture normally starts in service conditions. Thus containers used to contain beer and other drinks under pressure are often designed with bulges near the shoulder and the base. This prevents bottle to bottle contact in the central section of the wall. The area around the base of a jam jar is given a marked inward curve to protect the region in which thermal shock starts and thin-walled drinking glasses are bulged near the top to protect the rim (Fig. 81).

The second kind of evidence which provides useful information, particularly about the location of the fracture origin, is obtained by careful examination of the fracture surfaces. The most characteristic features of a fracture origin may be seen most readily by examination of the fracture surfaces of a glass rod or strip broken by bending (Fig. 82). There is a very smooth area immediately around the fracture origin. Around this there is a band which has a frosted or matt appearance and this in turn merges into a region which is macroscopically rough, often characterised by ridges radiating out

153

Fig. 81. Design of glass articles to protect vulnerable regions.

Fig. 82. Mirror area in the vicinity of a fracture origin.

from the fracture origin.   This ridged region is usually referred to as "hackle".   The increasing roughness of the fracture surface is an indication of increasing fracture velocity.
Forking of the fracture occurs in the region where the hackle
begins to form.   Whether or not one sees all these features on
the fractured rod surface depends on the strength of the rod.
If the rod has previously been considerably weakened by severe
abrasion, the fracture stress will be low and there will be no
forking.   The smooth mirror area may then extend over the whole
surface.   Laboratory experiments on the fracture of rod and

lath specimens which have been subjected to various degrees of
abrasion have shown that a simple relationship exists between
the radius of the mirror area, r, and the fracture stress.   It
has the form

$$\sigma_f = k \cdot r^{-\frac{1}{2}}$$

where k is a constant for a particular glass (Shand, 1959;
Kerper and Scuderi, 1964a; Orr, 1972).

The surface near the fracture origin of a bottle broken by
internal pressure shows exactly the same features as that seen
in the rods.   To recover the glass fragment containing the origin it is necessary to wrap the container with adhesive tape before carrying out the test.

Examination of smooth areas of the fracture surface somewhat remote from the fracture origin often reveals curves surface features called "rib marks" which give information about
the direction of propagation of the fracture (Fig. 83).   The

a.

b.

Fig. 83.   Rib marks on fracture surfaces.

direction is away from the centre of curvature and hence from
right to left in the photograph.   Each rib mark represents a
position of the crack front at some instant during its propagation.   It is clear from Fig. 83b that the fracture was moving
much more rapidly along the bottom surface than it was at the
top surface.   This is characteristic of a bending fracture, the
lower face in the photograph being that subjected to tension.

By marking directions of propagation of the fracture on different fragments from the same article, one can identify the region in which the fracture origin was located. More careful examination of fragments from this region may then result in the identification of the fracture origin.

It is not always easy to see the rib marks, especially on fractures which have propagated slowly. They are best seen by reflected light, but one may have to spend some time experimenting with different angles of illumination.

Even fainter surface markings called Wallner lines may be observed on some fracture surfaces. These represent loci of intersection between the crack front and acoustic waves propagating in the glass. From careful measurements of the curvature of these lines it is possible to calculate the variation of fracture velocity as the crack propagates. This is interesting, but not an especially useful part of a fracture analysis exercise. If one is particularly interested in measuring fracture velocities, this can be done more easily in other ways (Schardin, 1959).

The reader is referred to a number of excellent articles on fracture analysis (Preston, 1931, 1939; Murgatroyd, 1942; Orr, 1972) which are well illustrated by diagrams and photographs of fracture patterns and surface features.

CHAPTER VI

REFRACTIVE INDEX AND DISPERSION

A. Introduction

The optical properties of glasses are important in many applications, ranging from practical everyday use in windows and
glass containers through those that are more critical but still
familiar, e.g. camera lenses, to the most recent and revolution-
ary application in fibres for optical communications.

It requires some effort of the imagination to appreciate
the extent to which our knowledge of the physical world has been
gained by exploiting the optical properties of oxide glasses.
Astronomy has developed by the use of glass lenses and mirrors
in telescopes; biology, medicine and materials science by the
use of glass lenses in microscopes. It is also difficult to
imagine how the properties of the electron could have been studied and the electronics industry developed without the use of
transparent glass envelopes for vacuum systems, which greatly
facilitated those early observations leading to the discovery of
the electron and the first measurements of its properties.

The study of the optical properties of glasses, or expressing this in a more general way, the study of the interactions of
glasses with electromagnetic radiation, has proved of great value especially in recent years, for the investigation of glass
structure and of the nature of the chemical bonds in glasses.
An excellent account of this field has been given by Wong and
Angell (1976). This type of work involves the use of radiation
over a very wide wavelength range from X-rays to the very long
wavelength infra-red; and, of course, one may regard the study
of the A.C. electrical properties as a study of optical properties at even longer wavelengths.

The range of wavelengths which is important for practical
applications of glasses is much more limited; from about 200 nm
to 20 μm. (The visible range of the spectrum is from 400 nm to
750 nm.) All glasses that are good electrical insulators
transmit radiation to some degree. Oxide glasses free from
transition metal oxides and certain other colouring agents are
very transparent in the visible, near ultra-violet, and near
infra-red. Sulphide glasses and some other chalcogenide glasses

are less transparent in the visible but transmit to significantly longer wavelengths in the infra-red.   In the next chapter
an account will be given of the radiation-transmitting properties of glasses and how they are affected by glass composition
and the radiation wavelength in the range of practical interest.
The present chapter is devoted to an account of the refraction
of light by glass and of applications which depend on this phenomenon, namely the use of glasses in optical systems.

B.   Refractive Index

1.   Measurement

As explained in elementary texts, refraction of light rays
passing from one medium to another occurs because the velocity
of light is different in the two media.   When the first medium
is a vacuum the ratio $n = c/c_m$ is termed the refractive index
of the second medium.   c is the velocity of light in vacuo and
c_m that in the second medium.   At most wavelengths c is greater than c_m.   The refractive index is most conveniently measured
by application of Snell's law, $n = \sin i/\sin r$, where i and r
are respectively the angles of incidence and refraction at the
surface of the material.   For the obvious reason of practical
convenience, the measurements are nearly always carried out in
air, no correction usually being made for the fact that the refractive index of air is not exactly unity.

The refractive index of a glass, as of other materials,
varies with wavelength.   This variation can be considerable
and it is necessary for the application of glasses in optical
systems to have accurate information on the variation, not only
in the visible part of the spectrum but also, for some applications, in the near infra-red and ultra-violet.   The very demanding design specifications of modern lens systems require
that the properties of optical glasses be controlled with great
care.   Thus the refractive index at a specified wavelength of
any individual melt of a commercial optical glass must not differ by more than 0.001 from the value quoted for that type in
the manufacturer's catalogue.   The variation of index within
any one melt should not vary from the mean value by more than
$\pm 1 \times 10^{-4}$ and the manufacturer will usually be able to supply
individual blocks within which the maximum variation will be

less than ±1 x 10⁻⁶ (Schott Optical Glass Catalogue, 1972).
These close tolerances require the ability to make very precise
measurements of refractive index.

The more common methods used for measurement in the visible
spectrum are described in a number of text books on Optics (e.g.
Tenquist et al., 1970).   A light source giving one or more
sharp, easily distinguishable spectral lines is required.   This
is usually a gas or metal vapour discharge lamp (Section D.1).
When an accuracy of one or two units in the fourth decimal place
is sufficient, use of a commercial Pulfrich or Abbe refractometer is probably the most widely used method.   Both instruments
depend upon the measurement of the critical angle of refraction.
Figure 85 shows the paths of near-critical rays through a Pul-

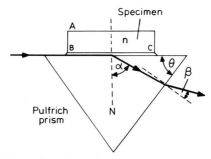

Fig. 85.   Pulfrich refractometer.

frich refractometer.   Preparation of the sample for measurement is relatively simple.   The glass block, which should be
at least 1 mm thick, is polished on faces AB and BC.   Face BC
should be flat to within one wavelength and the corner ABC
should be sharp, the angle being close to 90°.   The specimen
rests on the glass prism of the instrument.   The refractive
index of the prism, N, at the wavelength of the measurement
must be accurately known.   A very small quantity of a liquid
having a refractive index intermediate between the specimen and
the prism is used to form a thin liquid film between the surface BC and the upper face of the instrument prism.   The critically refracted ray is incident at 90° on the surface of the
prism.   It is refracted once at an angle α and emerges from
the second face at an angle β to the prism face normal.   Rays
incident at an angle less than 90° emerge below the critical
ray, but no rays can emerge above it because they cannot enter

the prism.   The measurement of angle β involves the alignment with a cross wire in the instrument telescope of a sharp boundary between the bright and dark sections of the field of view. This telescope is mounted on a circular vernier scale, which can be rotated about an axis normal to the plane of the diagram. By applying Snell's equation twice, once for each refraction, the following equation is obtained, relating n and the angle, β

$$n = \sin\theta \, (N^2 - \sin^2\beta)^{\frac{1}{2}} + \cos\theta.\sin\beta. \qquad (76)$$

If a higher accuracy is required, with results accurate to one or two units in the fifth or even the sixth decimal place, it is necessary to use either a high precision spectrometer with the specimen in the form of a relatively large, carefully made prism, or a V-block refractometer, e.g. of the Hilger-Chance type (Hughes, 1941; Holmes, 1945; Harper and Boulton, 1969). The specimen, S, is a prism (Fig. 86) the surfaces of which have

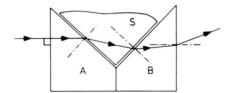

Fig. 86.   V-block refractometer.

been ground flat and at an angle close to 90° so that it is a good fit into the V formed by the faces of fixed prisms of the instrument A and B.   A liquid having a refractive index close to that of the specimen is used to give good optical contact between its surfaces and those of the instrument prisms.   Monochromatic light is directed by a collimator onto the vertical face of prism A and the angular deviation of the light passing through the composite prism is measured by means of a telescope mounted on a circular scale.   Knowing the refractive indices of prisms A and B, that of the specimen can be calculated.

For an account of other high precision methods of measuring refractive indices, the reader is referred to an article by Werner (1968).   This suggests that it should be possible to construct an instrument based on an interferometric principle capable of giving results accurate to one in the seventh decimal place.

The measurement of refractive indices outside the visible

range of wavelengths necessitates the use of special instruments
equipped with suitable photodetectors.

2. Effects of Glass Composition

When considering the effect of composition on the refractive
index of oxide glasses it has been usual to quote values at the
mean wavelength of the sodium doublet,n_D,i.e.589.3 nm. In recent
years this has been replaced as a standard wavelength by the
helium d line at 587.6 nm. The effect of glass composition on
dispersion, the variation of refractive index with wavelength,
is discussed later.

A great deal of information on the effect of composition is
available, both for simple compositions and for the more complex
optical glass compositions. Much of this has been summarised
by Morey (1954) and Mazurin et al., (1975). It is sufficient
here to present a very small selection of data to compare the
refractive indices of one component glasses and other simple
substances (Table XIV) and to show how the refractive index
varies with composition in a number of two- and three-component
glasses.

Vitreous beryllium fluoride has probably the lowest refractive index of the inorganic glasses. Most oxide glasses have
refractive indices between 1.45 and 2.0, but in recent years
new oxide glass-forming systems have been discovered containing
compositions for which the refractive indices are considerably
higher, e.g. the tellurite glasses with values as high as 2.3 -
2.4 (Stanworth, 1952). The chalcogenide glasses have higher
values still, but for these materials the refractive indices in
the infra-red are of greater interest because of their use as
infra-red transmitting windows. Values in the range 2.0 - 3.5
can be obtained, which may be compared with 4.1 for germanium,
another material fairly widely used as an infra-red window
material.

In many two-component glass systems, e.g. the alkali silicate systems (Fig. 87) and the alkali fluoride-beryllium fluoride systems (Fig. 88), the refractive index varies almost linearly with composition. However there are other systems in
which the composition-dependence shows interesting departures
from linearity. These have been interpreted in terms of changes

TABLE XIV

Refractive index n_d at $\lambda = 587.6$ nm of some one-component glasses and inorganic compounds

Glasses	n_d
BeF_2	1.275
B_2O_3	1.458
SiO_2	1.458
GeO_2	1.607

Crystals	
SiO_2: α-Quartz	1.544-1.553
Tridymite	1.469-1.473
Cristobalite	1.484-1.487
LiF	1.392
NaF	1.325
NaCl	1.550
KCl	1.490
KBr	1.560
MgO	1.737
Diamond	2.419

in the co-ordination number of the network-forming cations. Evstropiev and Ivanov (1963) suggest that, in the R_2O-GeO_2 glasses (Fig. 89), the first additions of alkali oxide result in some of the germanium atoms changing their oxygen-co-ordination from 4 to 6 and that this continues up to 15 mol.%, beyond which further addition of alkali results in a reversal of this process. The change in direction of the lines of equal refractive index in the $Na_2O-Al_2O_3-SiO_2$ system (Fig. 90) are also due, almost certainly, to a co-ordination number change. When the Na_2O/Al_2O_3 ratio is greater than 1, it is possible for all the Al^{3+} cations to be four co-ordinated by oxygen. All these ions may replace Si^{4+} cations in the glass network. To preserve the electrical charge balance it is necessary that the glass should contain a number of sodium ions at least equal to the number of Al^{3+} ions. When the Na_2O/Al_2O_3 ratio is less than 1, only a fraction of the Al^{3+} ions may replace the Si^{4+} ions.

3. Molar and Ionic Refractivities

When considering effects of composition on the refractive

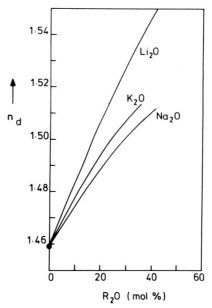

Fig. 87. Effect of composition on the refractive index of glasses in the systems R_2O-SiO_2 (Vogel and Gerth, 1958).

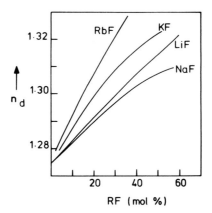

Fig. 88. Effect of composition on the refractive index of glasses in the systems RF-BeF_2. (Vogel and Gerth, 1958).

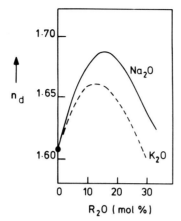

Fig. 89. Effect of composition on the refractive index of glasses in the systems R_2O-GeO_2.

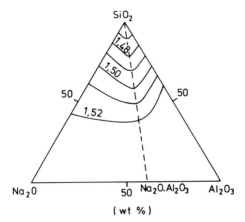

Fig. 90. Effect of composition on the refractive index of glasses in the system Na_2O-Al_2O_3-SiO_2.

index of inorganic solids, it can be more instructive to calculate, from the measured refractive index values, a quantity which is likely to depend in a simpler way than the refractive index itself on the polarizabilities of the constituent ions. The polarizability of a material, α, is a constant relating the intensity of electric polarization, P, to the field, E, which produces it, i.e.,

$$P = \alpha E$$

The polarizability, dielectric constant ϵ and refractive index at a particular frequency are related by

$$n^2 = \varepsilon = 1 + 4\pi\alpha \qquad (77)$$

If one seeks to relate these quantities to properties of the constituent atoms or ions, the most elementary theory which can be used as a basis for the relationships is that due to Lorentz (1906). This will be considered in more detail in a later section. Unit volume of the material is considered to contain n_0 independent oscillators each having a polarizability, α_0. The macroscopic polarizability is related to α_0 by

$$\alpha = n_0\alpha_0 / (1 - 4\pi n_0\alpha_0/3). \qquad (78)$$

On combining Equations (77) and (78) one obtains

$$(n^2-1)/(n^2+1) = 4\pi n_0\alpha_0/3 \qquad (79)$$

and on multiplying both sides by the molar volume, V_m,

$$V_m(n^2-1)/(n^2+1) = R_m = 4\pi n_0 V_m \alpha_0/3$$
$$= 4\pi N_m \alpha_0/3 \qquad (80)$$

R_m is called the molar refractivity of the material and is clearly related in a simple way to the polarizability of the atomic or ionic oscillators which the material contains. If the material contains more than one kind of ion, it seems reasonable to expect that the molar refractivity should be proportional to the sum of the products $p_i\alpha_i$ where p_i is the fraction of ions i in the material. This simple relationship is obeyed very well by the alkali halides (Fajans and Joos, 1924). Table XV lists the values for the *refractivities* of a number of ions, taken from Seitz (1940) and Scholze (1965). The relationship between the molar refractivity R_m of compound $X_m Y_n$ and the ionic refractivities of X and Y is simply

$$R_m = mR_X + nR_Y$$

Unfortunately the simple additive formula cannot be applied so easily to oxide glasses. In most glasses, the largest contribution to the molar refractivity of the glass comes from the oxygen ion. A few simple calculations are sufficient to show that the oxygen ion refractivity is not constant. It depends on whether the oxygen is non-bridging or bridging and it is also affected by the proximity of other cations. From the measured molar refractivity of 7.44 cm^3 of silica glass and using the value of 0.10 cm^3 from Table XV for the ionic refractivity of silicon, one can calculate the ionic refractivity of the oxygen ion as follows

$$R(\text{oxygen}) = (7.44-0.1)/2 = 3.67 \ cm^3.$$

TABLE XV

The refractivities of ions

Ion	R_I cm^3
F$^-$	2.5
Cl$^-$	9.00
Br$^-$	12.67
I$^-$	19.24
Li$^+$	0.20
Na$^+$	0.50
K$^+$	2.23
Cs$^+$	6.24
Be^{++}	0.10
Mg^{++}	0.28
Ca^{++}	1.33
Ba^{++}	4.30
Pb^{++}	3.10
B^{3+}	0.05
Al^{3+}	0.17
Si^{4+}	0.10
P^{5+}	0.07

However, if one calculates the oxygen ion refractivity using the experimental refractivity values for a number of crystalline silicates, much higher values are obtained, up to 4.77 cm^3. (Fajans and Kreidl, 1948). The effect of the proximity of other cations is shown in Table XVI. Thus the ionic refractivity values are

TABLE XVI

Refractivities of oxygen anions in crystalline orthosilicates

Material	$R_{I,O}$ cm^3
25% Na$_2$O.75% SiO$_2$ glass	4.72
Mg$_2$SiO$_4$ crystal	3.83
Ca$_2$SiO$_4$ crystal	4.53
Sr$_2$SiO$_4$ crystal	4.67
Ba$_2$SiO$_4$ crystal	4.97

of little practical value for calculating molar refractivities
of oxide glasses.   It is however of some practical value when
developing optical glasses to bear in mind the qualitative in-
dications, provided by the ionic refractivity values, of the
relative contributions of various ions to the molar refractivity.
Note however that the refractive index depends upon the molar
volume of the glass, V_m, as well as on the molar refractivity.
Thus equation (80) may be re-arranged as follows

$$n^2 = (1 + 2Y)/(1 - 2Y) \tag{81}$$

where $\qquad\qquad Y = 4\pi N_m \alpha_0 /3V_m.$

Hence a low value of V_m as well as a high value of R_m are both
factors which result in a high refractive index.

Because of this effect of density, it can be instructive
when studying the effect of composition on refractive index to
allow for the density effect by calculating the molar refractivity and plotting this against composition.   Figure 91 shows

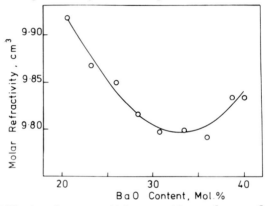

Fig. 91.   Effect of composition on the molar refractivity of
glasses in the system BaO-B_2O_3.   (Warner and Rawson, 1978).

this for the BaO-B_2O_3 system.   The presence of the marked minimum is not suggested by an examination of the refractive index
values themselves, which increase almost linearly with BaO content.   Presumably the minimum is associated with the well-known
changes in co-ordination number which result when a basic oxide
is added to B_2O_3.

Because of the great importance of optical glasses, a considerable amount of effort has been expended in attempts to de-
vise methods of calculating the refractive index from the glass

composition.   An excellent summary of this work and of the various formulae which have been developed has been given by Scholze
(1965, 1977).

Recently a more fundamental approach to the understanding
of the relationship between refractive index and composition has
been described by Pantelides and Harrison (1976).   They have
applied it with considerable success to the various forms of silica.   Their approach is a complicated one, involving the application of wave mechanics.   Any reader firmly resolved to understand refractive index and unafraid of atomic orbitals is likely
to find this paper of considerable interest.

4.   Effects of Heat Treatment

Reference was made in Chapter I to the effects on the properties of a glass of variations in heat treatment.   The viscosity results of Lillie described in Chapter II show viscosities
measured in the transformation range changing with time as the
configuration changes to that in equilibrium at the temperature
of measurement.

Another general feature of the properties of glass is that
measurements made below the transformation range, e.g. at room
temperature, are affected by the previous heat treatment, especially that given whilst the glass was in,or as it passed through
the transformation range.   Thus a glass which has been rapidly
cooled from a high temperature has a lower density than the same
material when cooled more slowly.   The more rapidly cooled glass
has a higher fictive temperature;   its configuration corresponds
to that in equilibrium at a higher temperature.   As has been
shown above, the refractive index of a material is affected by
its density.   It is not surprising therefore that this property
is also affected by cooling rate;   the higher the cooling rate
through the transformation range, the lower the refractive index.

Because of the importance of precise refractive index control to the optical glass industry a number of thorough studies
have been made on the effects of heat treatment on refractive
index.   The results of Lillie and Ritland (1954) shown in Fig.
92 and Fig. 93 are typical.   The results of Fig. 92 are room
temperature refractive index values for samples of an optical

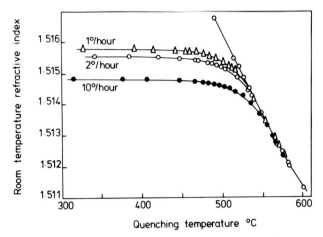

Fig. 92. Effect of quenching temperature on the room-temperature of refractive index of a borosilicate optical glass.

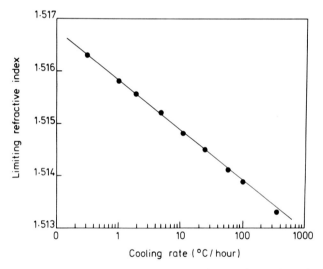

Fig. 93. Effect of cooling rate on the refractive index of samples cooled continuously to room temperature.

borosilicate glass which were cooled at different rates. Each point is for a sample which was quenched to room temperature by removing it from the furnace during the cooling period. The straight line on the right-hand side of the diagram is for specimens which were held at constant temperature for a time long enough to reach equilibrium and then quenched to room temperature. For the continuously cooled specimens, the slower the rate of cooling, the lower the temperature at which the glass

is able to attain the equilibrium configuration during the cooling period.

  The results in Fig. 93 show the effect on the room temperature refractive index of varying the cooling rate over a very
wide range. The effect is very large in relation to the degree of reproducibility demanded for high quality optical glass.
It is therefore necessary to control the heat treatment very
carefully indeed to ensure that different parts of the same
block of optical glass experience a very similar temperature
time schedule. The requirements for optical annealing to obtain a uniform refractive index through the material are much
more stringent than that of merely ensuring that the disanneal-
ing stress in the material is below a specified level. A very
low level of stress in optical glass is of course also necessary since stress affects the refractive index and makes the
material bi-refringent.

C. Dispersion

  In the following sections the phenomenon of dispersion and
its relationship to glass composition is discussed from a number
of points of view. A brief account is given of an elementary
theory of dispersion, which relates the dispersion in the visible
part of the spectrum to the presence of strong absorption bands
in the ultra-violet and infra-red. Results for glasses of simple composition are quoted which illustrate this dependence.
It is then possible to look at the compositions of commercial
optical glasses with some understanding of how their compositions affect their optical properties. Finally, a brief consideration of the problem of designing compound lenses to minimise abberations is used to explain why it is necessary to
manufacture such a wide range of optical glass compositions.

1. The Lorentz Theory of Dispersion

  The simple classical model of Lorentz (1906) gives an adequate qualitative picture of the phenomenon of dispersion. This
section describes the physical basis of the model and quotes
equations derived from it. Full accounts are given by Seitz
(1940) and Ditchburn (1963).

  The model pictures the material as consisting of electrons

bound to equilibrium positions by linear forces, i.e. if an electron is displaced from its equilibrium position by a distance
y, it experiences a force Ky acting to restore it to that position.   It is assumed that each electron is also subjected to a
damping force, which is proportional to its velocity, dy/dt.
If the electrons are subjected to the alternating electric field
of a light wave of amplitude E_O, they vibrate in response to the
field.   The equation of motion of each electron is therefore

$$m.d^2y/dt^2 + 2\pi m \gamma \, dy/dt + Ky + eE_O \exp(2\pi i \nu t) = 0 \qquad (82)$$

where ν is the frequency of the light wave;   e and m are respectively the electronic charge and mass and γ is the damping constant.   The solution of the equation is

$$y = - (e/4\pi^2 m).E_O \exp(i(2\pi\nu t-\phi))/((\nu_0^2-\nu^2)^2 + \gamma^2\nu^2)^{\frac{1}{2}} \qquad (83)$$

$$\text{where } \nu_0 = (K/4\pi^2 m)^{\frac{1}{2}} \text{ and } \phi = \tan^{-1}(\gamma\nu/(\nu_0^2-\nu^2))$$

Thus the electron displacement varies periodically, the phase
differing by ϕ from that of the light wave. ν_0 is the natural
frequency of oscillation of the electron.   The dielectric constant, ε, of a material containing n_0 such electrons per unit
volume is given by

$$\varepsilon = 1 + 4\pi\alpha = 1 + (n_0 e^2/m\pi)(\nu_0^2-\nu^2)/((\nu_0^2-\nu^2)^2 + \gamma^2\nu^2)$$

$$(84)$$

and the absorption coefficient η by

$$\eta = (2n_0 e^2/mc)(\gamma\nu^2/((\nu_0^2-\nu^2)^2 + \gamma^2\nu^2) \qquad (85)$$

where c is the velocity of light in vacuo.   The variations of
α and η with frequency are shown in Fig. 94.   The region of
frequencies which is of interest when considering the dispersion properties of optical glasses is that to the left of the
absorption maximum, i.e. on the long wavelength side of the absorption band.   At wavelengths well removed from the band, η
is low and $(\nu_0^2-\nu^2)^2$ is large compared with $\gamma^2\nu^2$.   The equation
for ε then simplifies to

$$\varepsilon = 1 + (n_0 e^2/m\pi)/(\nu_0^2-\nu^2) = n^2 . \qquad (86)$$

Thus the refractive index increases as the absorption band is
approached.   This is shown in Fig. 95 for a number of crystalline alkali halides, all of which have strong absorption bands

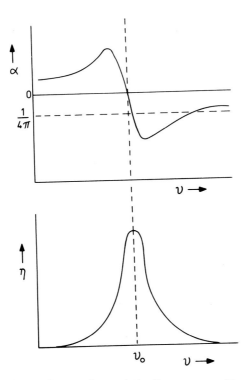

Fig. 94. Variation of α and η with frequency (Lorentz theory).

in the ultra-violet.   The nearer the band is to the visible part of the spectrum, the more marked is the dispersion.

Any real material has more than one strong absorption band and partly for that reason and partly because the Lorentz theory is an over-simplification, the variation of refractive index of transparent materials with frequency is more complicated than that given by Equation (86).   Various semi-empirical dispersion formulae have been proposed, e.g. the Helmholtz-Ketteler formula

$$n^2 - 1 = A + \Sigma_i \, c_i \lambda^2 / (\lambda_{0i}^2 - \lambda^2) \tag{87}$$

where the wavelengths λ_{0i} are taken to correspond to the centres of the strong absorption bands and the constants c_i are weighting factors related to their strengths.   Another formula is one that Kordes (1956, 1965) found to fit the experimental results for a number of transparent solids

$$1/(n^2-1)^2 = 1/(n_\infty^2-1)^2 - \lambda_s^2/\lambda^2 (n_\infty^2-1)^2 \; . \tag{88}$$

From the straight line obtained on plotting $1/(n^2-1)^2$ against

172

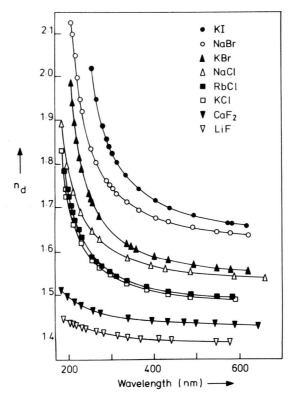

Fig. 95. Variation of refractive index with wavelength for various crystalline halides.

$1/\lambda^2$ it is possible to obtain the value of λ_s. The value obtained for each material was found to agree reasonably well with the wavelength of the first strong absorption in the UV. The main value of formulae of this kind, if they provide a good fit to the experimental data, is for interpolation purposes.

Although the Lorentz theory provides a satisfactory simple picture linking dispersion to the presence of absorption bands, the modern theory of the optical properties of solids is considerably more complex and is concerned with a much wider range of phenomena than the dispersion properties of optical glasses in the visible spectrum (Abeles, 1972).

2. The Relationship Between Dispersion and UV Absorption for Simple Glass Systems

The amount of detailed information on the dispersion prop-

erties of simple silicate and borate glasses is unfortunately
rather limited.   However it appears to be generally true that
mean dispersion n_F-n_C (see the following section) increases with
increasing basic oxide content of the glass (Figs. 96, 97 and
98).   For both alkali silicate and borate glasses there is à

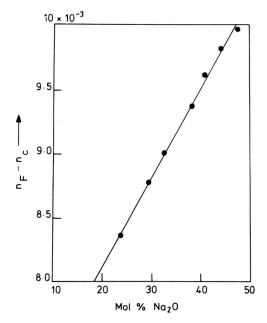

Fig. 96.   Variation of mean dispersion with composition of
glasses in the system Na_2O-SiO_2.

shift of the ultra-violet transmission cut-off to longer wavelengths as the alkali content increases (McSwain et al., 1963;
Hensler and Lell, 1969;   Sigel, 1973).   Thus for these systems
at least, there is the expected correlation between dispersion
and UV absorption.

The UV absorption is generally believed to involve the excitation of electrons associated with the oxygen anions in the
glass, and Stevels has suggested (1948, 1953) that the excitation energy will be less for the non-bridging oxygens produced
when basic oxides are introduced into the glass composition.
This explains the shift of the UV cut-off to longer wavelengths.
However it does not explain why the mean dispersion values of
lead silicate glasses (Fig. 97) are so high.   It is probable
that the high dispersion of these glasses is due to the presence

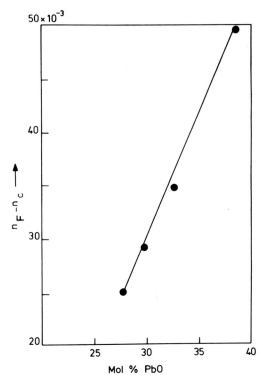

Fig. 97. Variation of mean dispersion with composition of glasses in the system PbO-SiO$_2$.

of a strong absorption in the near UV involving the excitation of electrons associated with the divalent lead ion (Paul, 1970; Stroud and Lell, 1971).

D.   Optical Glasses

1.   Terminology

In optical glass catalogues, values of refractive index are quoted at a number of wavelengths corresponding to lines in the emission spectra of certain elements.   These spectral lines are denoted by lower and upper case letter symbols (Table XVII) which are used as suffixes to denote the wavelength at which a property is specified.   Thus n_d represents the refractive index measured using the yellow helium line at 587.56 nm.   Dispersion values are quoted as differences between refractive indices at two specified wavelengths.   The mean dispersion, $n_F - n_C$ covers the wavelength range between the hydrogen C and F lines.   This amounts

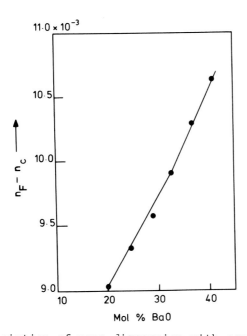

Fig. 98. Variation of mean dispersion with composition of glasses in the system BaO-B$_2$O$_3$ (Warner and Rawson, 1978).

TABLE XVII

Spectral lines commonly used for refractive index measurement.

Symbol	Wavelength nm	Element	Colour
i	365.01	Hg	UV
h	404.66	Hg	Violet
g	435.84	Hg	Blue
F$^{\prime}$	479.99	Cd	Blue
F	486.13	H	Blue
e	546.07	Hg	Green
d	587.56	He	Yellow
D	589.29	Na	Yellow
C$^{\prime}$	643.85	Cd	Red
C	656.27	H	Red
r	706.52	He	Red
s	852.11	Cs	Infra-red
t	1013.98	Hg	Infra-red

to most of the visible spectrum, including the important central region where the eye is most sensitive. The ratios obtained by dividing other differences, e.g. n_c-n_r or n_g-n_F by the mean dispersion are called partial dispersions, i.e.

$$P_{x,y} = (n_x-n_y)/(n_F-n_c).$$

The Abbe number, ν_d, is defined by the equation

$$\nu_d = (n_d-1)/(n_F-n_c).$$

When used in conjunction with n_d it is useful in classifying optical glasses (section D.3) and in carrying out elementary lens design calculations (section E).

2. A Brief History of the Development of Optical Glasses

When the first microscopes and telescopes were made, only one type of glass was available and it was impossible to make objective lenses free from chromatic abberation. Newton in his "Optics" published in 1704 inferred that dispersion is a property of light itself rather than the result of an interaction between the light and the medium through which it is transmitted. This opinion, based on experiments carried out using a number of transparent materials, led him to conclude that it was not possible to make an achromatic lens. However the development by Ravenscroft of lead crystal glass early in the eighteenth century soon changed the situation. This high lead content glass has a significantly higher refractive index and dispersion than the glasses previously known. This and similar glasses came to be known as "flint" glasses by opticians to distinguish them from the older "crown" glasses. (The two terms have persisted until the present day and can lead to confusion since it is common in the United States to refer to a colourless container glass as a flint glass.)

About 1730 it was discovered independently by a number of opticians that it was possible to considerably reduce chromatic abberation by using a combination of lenses; one made from crown glass and the other from the new flint. Once this discovery had been made, much effort was directed towards making more homogeneous optical glasses and making larger lenses for telescopes. A significant improvement in quality was made by a Swiss watchmaker, Guinand, who introduced the technique of mechanical stirring in the melting of optical glass. Much of the development of this process was carried out in collaboration

with the physicist Fraunhofer at a specially constructed glass
works in Bavaria.   Guinand's techniques were rapidly adopted
in France and England.

It is interesting to note the involvement of a number of
well-known scientists in the development of optical glasses.
Fraunhofer has just been mentioned.   In England Michael Faraday
was deeply involved for a number of years in experiments designed to make large melts of lead silicate glass of high quality
for telescope objects.   His account of this work (Faraday,
1830) is extremely interesting and gives a very good idea of the
careful experimental and largely empirical approach which one
has to follow even today when attempting to improve glass quality.

The next major advance occurred between 1874 and 1891 as a
result of the collaboration between Otto Schott, a chemist, and
Ernst Abbe, a physicist.   Their main objective was to develop
compositions with which it would be possible to design lens
doublets in which chromatic and spherical abberation could be
corrected to a much higher degree.   Schott made a very wide-
ranging study of the effects of glass composition on the optical
properties, introducing components which had not been previously
used, such as B_2O_3, BaO, F_2O_5, ZnO, Sb_2O_3 and various fluorides.
This work is of outstanding historical importance and represents
the first major scientific study of glass properties.

As significant as their scientific work, was the development by Schott and his colleagues of techniques for making the
new glasses on a commercial scale.   Another interesting feature
is the financial involvement of the Prussian government - perhaps the first example of state backing for an advanced technology; and a successful one too!   In 1886 the Jena Glass
Works, founded to make the new glasses, issued its first price
list.   This contained 44 compositions, 19 of which were essen-
tially new.

The reader interested in a more detailed account of the
development of this important sector of the glass industry
should consult the book by Douglas and Frank (1972) and articles
by Atma Ram (1961) and Deeg (1965).

3.   Compositions and Properties

The extent to which the introduction of Schott's new glass

compositions made possible the design of compound lenses of improved performance cannot be appreciated without some knowledge
of how limitations in the range of optical properties available
affect the ability of a lens designer to reduce abberations.
This is considered briefly in the next section.   However it is
easy to show how the available range of properties has expanded
over the 90 years since the publication of the first Schott catalogue.   Figure 99 is a way of displaying the properties of

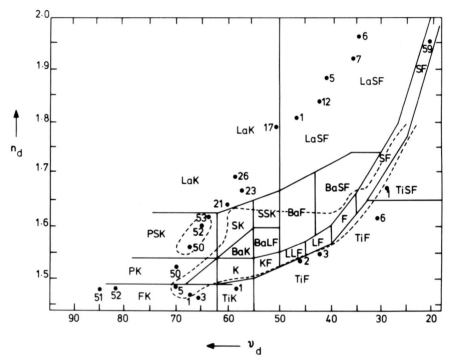

Fig. 99. A chart showing the range of refractive index and
Abbe number values available in commercial optical glasses.

optical glasses in a way which is convenient to the lens designer.
The glasses available in 1870 fall within the two areas bounded
by dotted lines.

The number of optical glasses now listed in the catalogue
of a major manufacturer is typically about 200.   Current chemical compositions are not published but it is possible to obtain
from the literature some indication of the compositions which
have been used to give optical properties similar to those of
corresponding present day compositions.   The most complete in-

formation is available for glasses which have been listed in
Schott catalogues.   A selection is given in Table XVIII.   The
combination of letters in the table, e.g. FK, SF, etc., refers
to the area in which the glass is to be found in Fig. 99.

When discussing the properties it is more convenient to refer to Fig. 100, which shows the relationship between the mean

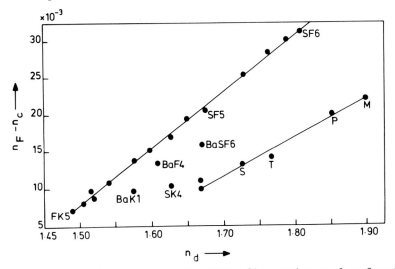

Fig. 100.   Relationship between mean dispersion and refractive
index for commercial optical glasses.

dispersion and the refractive index.   There is a simple, almost
linear relationship for the series of crown and flint glasses
containing the groups K, LF, F and SF (the upper line in the
figure).   In this series the major change in composition is an
increase in PbO content at the expense of other constituents.
High refractive index glasses are made possible in this series
because glasses of high PbO content can be made which are relatively stable and easy to melt.   The high refractive index is
due to the high refractivity of the Pb^{2+} ion.   This also causes
a high dispersion because of the strong absorption in the near
UV caused by this ion.

One of Schott's major innovations was the use of BaO instead
of PbO to make high refractive index glasses.   This gives rise
to the groups BaK, SK, BaF and BaSF.   Since there is no strong
absorption in the near UV associated with the Ba^{2+} ion, a high
refractive index is obtained without at the same time increasing
the dispersion.

TABLE XVIII

Composition and Properties of Some Commercial Optical Glasses

Type Number	n_D	n_F-n_C	v_d	SiO_2	B_2O_3	Al_2O_3	Na_2O	K_2O	CaO	ZnO	BaO	PbO	TiO_2	KHF_2	As_2O_3	Sb_2O_3	Source
FK3	1.4645	0.007063	65.77	47.7	17.4	14.0	2.2	2.4						16.0	0.3		H
FK5	1.4875	0.006924	70.41	56.9	15.7			5.6						21.6	0.2		H
BK1	1.51009	0.008038	63.46	71.4	6.5		5.2	13.9	2.0						1.0		H
K3	1.5185	0.0087	58.8	68.3	2.2	0.2	14.4	7.0	4.6			2.9			0.4		J
KF3	1.5145	0.009406	54.7	70.0		1.5	16.6			5.0		3.5	1.2	1.2	1.0	0.5	W
LLF2	1.5407	0.011464	47.17	63.2			5.0	8.0				23.5			0.3		H
LF7	1.57501	0.013860	41.49	53.9			2.5	7.9				34.9			0.8		H
F8	1.5955	0.015200	39.18	50.2	0.4		3.8	5.6				39.7			0.3		H
F2	1.62004	0.017050	36.37	45.7			3.6	5.0				45.1			0.6		H
SF2	1.64769	0.019135	33.85	40.9			0.5	6.8				50.8			1.0		H
SF5	1.6727	0.020884	32.21	38.7			1.5	3.9				55.6			0.3		H
SF10	1.72825	0.025634	28.41	35.3			2.0	2.5				55.7	4.0		0.5		H
SF14	1.76182	0.028718	26.53	32.1		1.2	1.0	1.0				60.5	3.7		0.5		H
SF11	1.78472	0.030468	25.76	29.2		2.5	0.5					63.3	4.0		0.5		H
SF6	1.80518	0.03166	25.43	26.9		0.5	1.0					71.3			0.3		H
BaK1	1.57250	0.09948	57.55	47.7		1.0	1.0	1.0		8.6	29.0				1.0		H
SK5	1.589130	0.009615	61.21	38.7		5.0					40.1				1.0	0.3	H
SK1	1.610250	0.010761	56.71	40.1	5.7	2.5		1.0		8.5	42.2	0.5			0.5		H
SK3	1.60881	0.010332	58.92	35.0	11.9	4.5	0.5				45.9	0.6			1.0	1.6	H
SK4	1.62720	0.01045	58.63	33.2	1.9	5.5					47.8	0.3			1.0	0.3	H
SK10	1.622801	0.010945	56.9	30.6	11.7	5.0	0.1			2.0	48.2	0.7			1.0	0.8	H

TABLE XVIII (continued)

Type Number	n_D	$n_F - n_C$	ν_d	SiO_2	B_2O_3	Al_2O_3	Na_2O	K_2O	CaO	ZnO	BaO	PbO	TiO_2	KHF_2	As_2O_3	Sb_2O_3	Source
BaF4	1.605620	0.013787	43.93	45.5			0.5	7.3		8.0	15.8	22.5			0.4		H
BaF9	1.643277	0.01341	47.96	35.3	5.3	2.2				8.2	36.1	12.3	0.2		0.4		H
BaSF2	1.66446	0.01854	35.83	38.3			2.5	4.4		5.0	13.0	34.5	2.0		0.3		H
BaSF6	1.667551	0.015921	41.93	34.9	4.5	0.5	1.1	1.9	5.6	4.6	25.8	18.0	2.5		0.3	0.3	W

H. "Optical Glass Manufacturing at Schott & Gen. Jena" CIOS Report Item 9 File XXXII-22

W. "Schott & Gen. Jena" CIOS Report Item 9 File XXXIII-69

J. "Japanese Optics" BIOS Report/JAP/PR/1308

The lower straight line in Fig. 100 and the data in Table XIX give the properties and compositions of high refractive in-

TABLE XIX

Rare Earth Containing Glasses (Meinecke, 1959)

Type	n_D	n_F-n_C	ν_D	B_2O_3	BaO	SrO	La_2O_3	Ta_2O_5	ThO_2
					weight per cent				
O	1.658	.011325	58.1	40	20	20	20		
N	1.686	.0118276	58.0	40	20		20		20
S	1.723	.013364	54.1	40			60		
T	1.767	.014922	51.4	26			33		41
P	1.850	.02024	42	20			36	28	16
M	1.898	.022676	39.6	16.6			37.5	29.2	16.7

dex glasses based on the $BaO-La_2O_3-B_2O_3$ system (Brewster et al., 1947; Hamilton et al., 1948). These glasses were discovered by Morey (1937) and were developed by the Kodak Company of America. Schott glasses of this type are in the areas LaK, LaF and LaSF in Fig. 99. The very low dispersion compared with the PbO-containing glasses is again due, presumably, to the absence of strong absorptions in the near UV. These glasses are of great value in the design of high performance camera lenses. They facilitate the design of lenses of high numerical aperture and low spherical aberration (Morian, 1973; Meinecke, 1959).

In this book we can consider only the major groups of optical glasses. The discussion has been limited to the effects of composition on refractive index and mean dispersion. However one must at least mention the fact that it is sometimes necessary to provide glasses which, for a given value of mean dispersion, have either particularly high or particularly low partial dispersions at the red or blue ends of the spectrum. These glasses can be produced by adding other constituents, e.g. fluorides, P_2O_5, ZnO or TiO_2. A brief discussion of the relationship between partial dispersion and composition has been given by Izumitani and Nakagawa (1965).

E. The Significance of Optical Properties in Lens Design

In this section a number of formulae derived from the simple thin lens formula are used to show how lens design has been affected by the availability of a wide range of optical glasses.

This is a superficial foray into a very complicated subject:
"real life" lens design is a complicated business and nowadays
makes considerable use of digital computers.

The thin lens formula giving the focal length in terms of
the refractive index of the glass and the radii of curvature of
the lens surfaces is

$$1/f = (n-1)(1/R_1 - 1/R_2) = K(n-1) \qquad (89)$$

The focal length f_T of a doublet, i.e. two simple lenses mounted
together is given by

$$1/f_T = 1/f_1 + 1/f_2 = K_1(n-1) + K_2(n_2-1)$$

where f_1 and f_2 are the focal lengths of the constituent lenses.

If the doublet is to have the same focal length for the C
and F wavelengths, i.e. $f_{TC} = f_{TF}$, the following equation must
be satisfied

$$K_1(n_{1C}-1) + K_2(n_{2C}-1) = K_1(n_{1F}-1) + K_2(n_{2F}-1) \ .$$

By re-arranging this equation and eliminating K_1 and K_2 by making
use of the formulae for the focal length of each lens for the
helium d line, it is easy to show that the condition for achro-
matism is

$$\nu_1 . f_{1d} + \nu_2 . f_{2d} = 0 \ . \qquad (90)$$

From this equation one begins to appreciate the significance of
the Abbe number.

Using the equations given above, one can easily show that
the more the two glasses differ in Abbe number, the smaller will
be the K values of the component lenses. This means that the
radii of curvature of the lens surfaces are reduced, with a consequent easing of the task of reducing other lens aberrations.

For the optical glasses which were available prior to
Schott's work, the mean dispersion increased linearly with refractive index. The properties of all the known glasses lay
close to a line having the equation

$$n_F - n_C = 0.07812 \, n_d - 0.10962 \ .$$

Thus, as the refractive index increased from 1.5 to 1.7, the
Abbe number decreased from 60 to 30. This imposed serious re-
strictions on the design of camera lenses, for example. For
such a lens to produce a flat stigmatic image the so-called
Petzval condition must be satisfied

$$n_{1d} . f_{1d} + n_{2d} . f_{2d} = 0 \ . \qquad (91)$$

Equation (90) must also be satisfied if the lens is to be achromatic for the C and F wavelengths. Both equations can be satisfied simultaneously only if $n_{1d}/\nu_{1d} = n_{2d}/\nu_{2d}$. This became possible only when the new glasses were introduced.

Finally, another important feature of the new glasses was that compositions were introduced for which partial dispersions could be varied independently (to some extent) of the Abbe number. For the old glasses the various partial dispersions varied linearly with Abbe number, e.g.

$$P_{C,t} = 0.5450 + 0.004743\ \nu_d$$
$$P_{F,e} = 0.4884 - 0.000526\ \nu_d$$
$$P_{g,F} = 0.6438 - 0.001682\ \nu_d$$

The new glasses showed deviations from these equations, some deviations being positive and others negative. By using at least one "abnormal" glass in designing an achromat, the focal length could be maintained constant over a wider region of the spectrum.

One can show from the equations already given that for a doublet which is achromatic for the C and F wavelengths, the focal lengths will differ for two other wavelengths X and Y by an amount related to the difference between the partial dispersions of the two glasses for these wavelengths. The relationship is

$$1/f_{Tx} - 1/f_{Ty} = (1/f_{Td})(P_{1,x,y}\ P_{2,x,y})/(\nu_1 - \nu_2) \qquad (92)$$

Using the old "normal" glasses, the difference between the partial dispersions is always the same, because of the linear relationships between the partial dispersions and the Abbe number. The problem of surmounting this limitation, and by so doing reducing the residual chromatic aberration was the main factor which gave rise to the collaboration of Schott and Abbe.

Figure 101 shows the variation of focal length with wavelength for two achromatic doublets both having the same f_d of 100 mm. One is designed with the normal glasses K7 and F8 and the other with "abnormal" glasses LgSK2 and KzFSN4. The reduction in secondary spectrum is clearly considerable.

F. Glass Optical Fibres

1. Established Applications

The use of glass and polymer fibres as flexible light guides has been well established in engineering and medicine for many

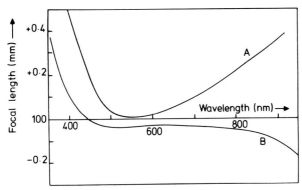

Fig. 101. The variation of focal length with wavelength for
two achromats.   A: using normal glasses K7 and F8.
B: using abnormal glasses LgSK2 and KzFSN4.

years (Kapany, 1960, 1967; Allan, 1973).   A ray of light directed into a fibre at one end in a direction making a small angle
with the fibre axis is propagated along the fibre undergoing
total internal reflections at the glass-air interface (Fig. 102).

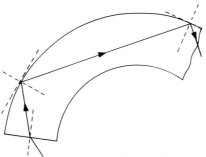

Fig. 102. Propagation of a light ray along an unclad glass
fibre.

The number of reflections depends upon the angle the incident
ray makes with the axis of the fibre and whether the fibre is
straight or not.   In a single component fibre of the type
shown in the figure, a small fraction of the light energy enters
the air on reflection at the surface.   Unless this surface is
very clean and smooth a significant energy loss occurs at each
reflection.   To eliminate this, optical fibres are invariably
made from two materials of differing refractive index.   The core
glass, in which the light propagates, is surrounded by a cladding made from a glass of lower refractive index.   The composite
fibre is made by placing a rod of core glass inside a tube of

cladding glass, the rod being a close fit inside the tube. This assembly is slowly fed in through the top of a vertical tube furnace inside which the rod and tube fuse together. The fibre is drawn downwards from the lower end of the composite rod and is wound continuously onto a motor-driven drum. Bundles of fibres, clad in plastic are normally used rather than individual fibres. These have a wide range of applications, including illuminators for stereoscopic microscopes, road signs, and dashboard signal systems for the detection of failed car lights. In a "coherent" light guide, used in engineering and medicine for examining otherwise inaccessible regions, an individual fibre must occupy the same relative position in the bundle at each end of the guide. The quality of the image and the size of the region which can be viewed is determined by the diameter of each fibre and the number of fibres in the guide.

In a number of devices manufactured by the electronics industry, fibre optic plates are used. These are made by stacking short fibres side by side and fusing them together under pressure. Large plates are used in some cathode ray tubes. The plate transfers the image from the fluorescent powder coated on the internal surface. Obviously such a plate must be vacuum tight, a difficult requirement to meet considering the very large number of fibres which must be sealed together.

Smaller plates are used in image intensifiers (Graf and Polaert, 1973). A special plate used in some of these devices, called a channel multiplier, is made from fibres the core glass of which is acid soluble. After the plate has been made, the core glass is dissolved away. The plate now consists of an assembly of very fine parallel holes. This is next heated in a reducing atmosphere. A conducting layer is produced in the surface of the cladding glass, which contains a proportion of easily reducible oxides such as PbO. This conducting coating is used to control the voltage gradient along the internal surface of the tubes when a potential is applied between electrodes applied to each surface of the plate. In use, the plate is mounted in a vacuum device. An electron entering the end of one channel is accelerated along the channel, following a path in which it makes a series of impacts with the channel wall.

On each impact more than one electron may be emitted from the
channel surface.   Thus for each electron entering one end of
the channel, many leave at the other.   This effect is of great
value for amplifying very faint light signals, the incoming
light being used first to produce photo-electrons from a layer
of a emitter material, these electrons being subsequently increased in number by the channel plate.   The final image is
produced by the electrons striking a fluorescent screen.
(Eschard and Manley, 1971;   Trap 1971;   Washington et al. 1971;
Mackenzie, 1974).

2.   Optical Communications

   A paper by Kao and Hockham, published in 1966 proposing the
use of glass fibres for transmitting information as modulated
light signals, initiated what has probably been the largest and
most intensive programme of research and development on glasses
that has ever taken place.
   Figure 103 shows the principle of a fibre optic communica-

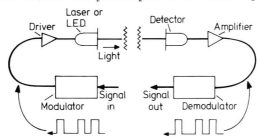

Fig. 103.   Fibre optic communication system.

tion system.   The input is shown as a series of electrical
pulses, which may have originated from a computer or have been
produced by pulse code modulation from speech in a telephone
system.   The electrical pulses are converted into a similar
train of light pulses by a light-emitting diode or a semi-conductor injection laser attached to one end of the fibre.   Some
loss of light occurs in the fibre and the pulse shape may also
be distorted.   It is therefore necessary to insert repeater
units periodically in the line to amplify and reshape the pulses.
At the receiving end, the signal is finally reconverted into
whatever form is required - a speech signal in the case of a
telephone system.

The main advantage of the system is the very high informa-
tion-carrying capacity of each link. There are also a number
of other advantages such as immunity from electrical interference and reduction of weight (an important factor in military
aircraft). For such a system to be sufficiently attractive for
use in a telephone network it is necessary that the number of
repeater units should be kept to a minimum. Cost analysis has
shown that they must be several kilometres apart. This depends
on being able to make fibres in which the light losses are reduced to a very low level. The optical glasses available in
1966 were not satisfactory. Losses, due mainly to absorption
by metal oxide impurities in the glass were of the order of 100
times too high. The main objective initially was therefore to
make glasses substantially free from these impurities. A second important objective was to make fibres with a controlled
radial distribution of refractive index since this was found to
be an important factor affecting the distortion of the light
pulses. These objectives have been largely attained and field
trials of fibre optic links are now being carried out in a number of countries with a high probability that such links will
come into widespread use in the early 1980's.

The literature in this field is extensive as consultation
of several excellent reviews will show (Miller et al., 1973;
Maurer, 1976, 1977; Barnoski, 1976; Gliemeroth et al., 1976;
Gossink, 1977). The 1976 review by Maurer is particularly
useful for its account of loss mechanisms in glasses, whilst
Gossink's article contains more information on the methods used
for making very high purity glasses.

Although the glass fibre is an essential component of the
system, the concept would not have been proposed had it not been
for the availability of semi-conductor light sources and detec-
tors, the former being capable of producing extremely rapid
changes in light intensity and the latter capable of producing
equally rapid changes in electrical output in response to a
modulated light signal.

The light absorbing impurities which must be eliminated are
oxides of the transition elements - iron, copper, vanadium,
chromium, manganese, cobalt and nickel. The colours due to
these oxides are discussed in detail in the following chapter.

For most of them the maximum permissible concentration is considerably less than 1 ppm.   Another troublesome impurity is
found in the glass in the form of hydroxyl anions.   These occur
bonded to silicons in the glass network.   The main hydroxyl
absorption is near 2800 nm but overtones of the absorption are
found near 950 nm, close to the wavelength region used in the
system (800-1000 nm).   The hydroxyl impurity originates from
water present in the raw materials and the atmosphere in which
the glass is produced.

Another potential source of energy loss is spatial variation in the refractive index of the glass.   Such a spatial variation is an intrinsic property of any transparent material,
even of a liquid such as water.   The scale of the variation is
usually significantly less than the wavelength of visible light.
It gives rise to scattering, the magnitude of which is inversely
proportional to the fourth power of the wavelength (Rayleigh
scattering).   Fortunately there is little difficulty in finding glass compositions in which the Rayleigh scattering is neg-
ligibly low in the near infra-red.   Figure 104 shows the varia-

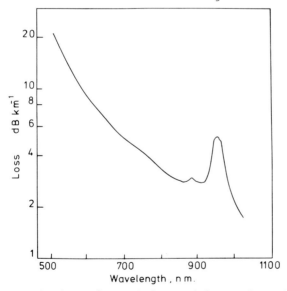

Fig. 104. Variation of total loss with wavelength for a
communications fibre (Gliemeroth et al., 1976).

tion of the total loss of a fibre (absorption plus scattering)
with wavelength.

The lowest losses have been achieved for fibres made from
glasses of high silica content by methods differing considerably from normal glass melting procedures.   As in conventional
fibre optics, the fibre consists of two glasses, the core having
the higher refractive index.   The method of manufacture is illustrated in Fig. 105.   The core material is deposited from the

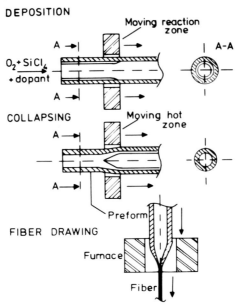

Fig. 105. Stages in the manufacture of a high silica glass
communications fibre (Cossink, 1977).

vapour phase on the inner surface of a silica glass tube.   Volatile compounds (usually chlorides) of the glass constituents
are carried in an oxygen gas stream into the tube, which is ex-
ternally heated.   The chlorides react to form oxides which may
deposit as a compact layer on the tube wall or may form an oxide
smoke in the tube.   If a smoke forms, this must be allowed to
settle on the wall and then be heated to a sufficiently high
temperature to fuse it.   The tube is subsequently collapsed to
a composite rod which is finally drawn to a fibre.   Elements
used to control the composition and hence the refractive index
of the core glass include titanium, germanium, boron and phos-
phorus.   Volatile compounds of these elements are converted
into their oxides in the process and form a homogeneous glass
with the main constituent, SiO_2.   In some types of fibre the

purity of the cladding is critical.   Fortunately, methods of
making very high purity silica, again using volatile silicon
compounds, have been in use for some time to meet the requirements of the semi-conductor industry.

Low loss fibres suitable for some communication applications have also been made from glasses prepared by conventional
methods, i.e. by melting a mixture of solid chemicals.   Very
high purity raw materials are required and it sometimes is also
necessary to use special melting techniques (e.g. direct induc-
tion melting in a cooled silica crucible) to prevent contamination of the melt by the crucible material.

The importance of the radial variation of refractive index
in the fibre was mentioned earlier in this section.   When the
development started, two types of fibre were considered.   These
are shown in Fig. 106.   The fibre diameter is of the order of

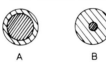

A   B

Fig. 106.   Types of communications fibre;   A: multimode,
B: single mode.

100 μm for both types.   In the multimode fibre the core diameter is about 60 μm but only about 5 μm in the single mode fibre.
The multimode fibre was designed for use in conjunction with an
incoherent L.E.D. light source, and the single mode fibre with
an injection laser.   The single mode fibre/laser combination
should have the higher performance but presents the more technical difficulties.   It was subsequently realised that a significant improvement in the performance of the multimode fibre from
the point of view of reduced pulse distortion could be achieved
if the core could be made with a parabolic radial refractive index variation (Fig. 107).   This reduces the difference between
the transit time along the fibre of on the one hand, a ray propagated along the fibre axis, and on the other rays inclined to
the axis.

Fibres with this type of refractive index profile are now
available.   When using the vapour phase technique, this is
achieved by a controlled variation in the mixture of volatile
compounds fed into the tube.

192

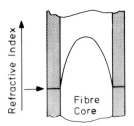

Fig. 107. Parabolic radial variation of refractive index in the fibre core.

CHAPTER VII

THE ABSORPTION OF RADIATION BY GLASSES

A.   Introduction

Mention has been made in the previous chapter of the practical significance of the fact that glasses are transparent to
electromagnetic radiation in some parts of the spectrum.   For
each glass the transmission varies with wavelength.   Depending
on the wavelength range, this variation may be determined either
by the major constituents of the glass composition or by constituents present in relatively small proportions and which are
deliberately added to control the transmission.

In the visible region of the spectrum the eye senses the
variation of transmission with wavelength, recognizing it as
colour.   Coloured glasses have been made from antiquity.   The
glass makers of the ancient world knew what minerals should be
added to their glasses and enamels to produce particular colours,
although no doubt they were not always successful in producing
the particular shade or intensity at which they were aiming.
(Their successors in the twentieth century sometimes have similar problems.)

The demands made on the glass industry to produce glasses
of controlled transmission are very varied in nature.   Some
that may seem trivial are nevertheless of considerable commercial importance.   It is quite common, for example, for companies supplying foods and drinks in glass containers to insist
either on a complete absence of colour or on some particular
tint of "colourless" glass.   They maintain, no doubt with justification, that this affects the customer-appeal of their product.   A less subjective requirement is for the colours of signal and warning lights used in various forms of transport to be
controlled.   For these applications various official colour
specifications exist which the glass has to meet (Wright, 1964).

Outside the visible spectrum, the requirements tend to be
of a more technical nature.   Thus glasses are required to
transmit ultraviolet radiation of wavelengths which can kill
bacteria or can excite phosphors to emit light in some types of
lamp.   Heat-seeking missiles use infra-red radiation emitted
from the target at wavelengths for which the atmosphere is

transparent.   This has led to much work on chalcogenide glasses capable of transmitting at wavelengths beyond 10 μm.   Infra-red transmitting silicate glasses are needed for heat lamps used for various applications including relieving muscular pains, drying paint, and rearing pigs.   These glasses are often made to be strongly absorbent in the visible to reduce glare from the lamp filament.

The radiation energy absorbed is usually dissipated as heat in the material.   In one application, the manufacture of reed switches, advantage is taken of this (Hoogendoorn and Sunners, 1969).   A reed switch is a glass-envelope device now used in many telephone exchanges.   Metal leads are sealed through the envelope to connect with the switch contacts.   The glass-to-metal seals are made by heating the glass by focussed radi-ation from a high intensity lamp, the glass being one specially formulated to absorb the lamp radiation.   This method provides a much cleaner atmosphere than flame sealing and the risk of oxidizing the switch electrodes is consequently reduced.   The process of heat transfer in glass furnaces, and in glass man-ufacturing generally, depends critically on the ability of the glass to absorb and emit infra-red radiation.

Less often, the absorbed radiation is re-emitted as radi-ation of longer wavelengths, giving rise to the phenomena of fluorescence and phosphorescence.   Fluorescence in oxide glass-es has been known for many years but has taken on a much greater significance with the development of the laser.   Laser glasses have been developed which are now being used in small pieces in laser range finders and in very large pieces in laser fusion experiments.

This long catalogue of practical applications, which is by no means complete, gives more than enough reason for taking an interest in the absorption of radiation by glasses.   However this is only half the story.   To a physicist or chemist the absorption and emission of radiation by a material signifies spectroscopy, a branch of science which for many years has con-tributed greatly to an understanding of the structure of matter, including glasses.

Although the emphasis of this book is primarily technolog-ical, without an understanding of principles technology would

be entirely empirical. This chapter will therefore make at
least some reference to what is known about the mechanisms by
which radiation is absorbed. The reader requiring more information on the spectroscopy of glasses is referred to the compre-
hensive account by Wong and Angell (1976).

B. Units and Nomenclature

The differential equation which describes the rate of change
of the intensity, I, due to absorption of an electromagnetic wave
as it propagates in a material in the positive x direction is
$dI/dx = - \gamma I$. The absorption coefficient γ is a property of
the material and depends on the wavelength of the radiation.
Integration of this equation gives $I = I_0 \exp(- \gamma x)$, usually
known as Beer's Law. I_0 is the intensity of the radiation at
x = O, this plane being taken to be that at which the radiation
enters the specimen. For coloured glasses the value of γ is
usually directly proportional to the concentration, c, of the
colourant material in the glass, i.e. $\gamma_\lambda = \epsilon_\lambda c$. It is the con-
vention to specify c in mole l^{-1}, and x in cm, whence the units
of ϵ_λ are litre mole^{-1} cm^{-1}. ϵ_λ is called the molar absorptivity or molar absorption coefficient. Some authors prefer
to quote their results using Beer's Law expressed in terms of
the base 10, i.e. $I = I_0 . 10^{-\alpha_\lambda x}$. α_λ is termed the extinction
coefficient. When using Beer's Law to record the results of
experimental measurements, the distance x is the thickness of
the specimen. Other commonly used terms are the absorbance
$(\ln(I_0/I))$ and the optical density $(\log_{10}(I_0/I))$. These must
be referred to a specific thickness of material.

The measurement of absorption by a specimen can be conveniently carried out only by comparing the intensity of the radiation leaving the specimen with that entering it. At each
interface a fraction, R, of the radiation intensity is back reflected and this must be taken into account when calculating
γ_λ or α_λ. Thus if I_{in} is the intensity just before the radiation enters the glass and I_{ex} is the intensity just after the
radiation enters the air again

$$I_{ex} = I_{in}(1-R)^2 \exp(- \gamma_\lambda x) \qquad (92)$$

The value of R depends upon the refractive index of the glass.
For normal incidence, R is given by

$$R = (n-1)^2/(n+1)^2 \ . \tag{93}$$

For a glass of refractive index n = 1.5, R = 0.04.

The development of optical communications using glass fibres has made it necessary for the glass technologist to become familiar with yet another way of expressing the attenuation of radiation intensity. The intensity reduction in the fibre is expressed in terms of decibels per kilometre ($dBkm^{-1}$). The intensities I_{ex} and I_{in} are said to differ by n decibels if n = 10 $\log_{10}(I_{in}/I_{ex})$.

The process of absorption of radiation by a material is one in which a quantum of radiation excites an absorption centre in the material from an energy level E_1 to a higher level E_2. $E_2-E_1 = h\nu$ where ν is the frequency of the radiation and h is Planck's constant. When the purpose of a study of radiation absorption by glass is scientific rather than technological it is understandable that the results are not presented as showing the variations of γ_λ or α_λ with wavelength but as their variation with a quantity more closely related to the energy difference E_2-E_1. Most frequently, spectra show the variation with the reciprocal of wavelength but occasionally energy units are used (electron volts). Table XX gives conversion factors for

TABLE XX

Energy Units in Spectroscopy

	cm^{-1}	eV	erg
cm^{-1}	1	1.24×10^{-4}	1.99×10^{-16}
eV	8066	1	1.60×10^{-12}
erg	5.03×10^{15}	6.24×10^{11}	1

relating one type of unit to another. It is sometimes helpful to be able to relate energy differences measured spectroscopically to the thermal energy in a material, expressed by kT where k is Boltzmann's constant and T is the temperature in °K. At room temperature (300°K) kT = 208.5 cm^{-1} = 0.0259 eV. This corresponds to a wavelength in the far infra-red.

C. Transition Metal Ion Colours

Most of the oxides of the first series of transition metals (titanium to copper) produce a strong colour when present in an

oxide glass at a concentration of less than one per cent. The
transition metal ions are incorporated in the glass structure,
each being surrounded by oxygen anions. The colour is due to
the excitation of electrons in the incomplete 3d shell of the
ion to higher energy levels. The energy differences between
the ground state of the ion and its various excited states are
affected by the electric field which the ion experiences due to
the surrounding oxygen ions (the ligand field). This field
varies according to the number and the type of oxygen ions in
the first co-ordination sphere and this in turn depends upon
the glass composition. Consequently the colour produced by a
given transition metal ion is affected by the composition.

The range of colours that can be produced by the transi-
Another important factor affecting the colour arises from
the fact that the transition metals are multivalent. For example, iron reacts chemically to produce ferrous and ferric compounds, containing the ions Fe^{2+} and Fe^{3+} respectively, whilst
vanadium has at least four valence states. In a glass containing one of these elements there will usually be two valence
states present. Given a sufficient length of time at the melting temperature, the relative concentrations of the two valence
states eventually reach an equilibrium. For example, if ferric
oxide is added to the batch of a soda-lime-silica glass and the
batch is melted to a glass which is held sufficiently long at
1500°C in contact with air, approximately 20% of the iron will
be present as Fe^{2+} ions and the remainder as Fe^{3+}. The equilibrium ratio is affected by the composition of the glass, the
melting temperature and the melting atmosphere. Since the different valence states produce different colours, the glass colour will change if any of these controlling factors is changed.

The range of colours that can be produced by the transition metals is considerable, as one can readily imagine from
this brief account. No doubt this is a very happy state of
affairs so far as the glass artist or craftsman is concerned.
He can use to advantage the subtle variations in tint that are
possible. To the technologist, required to melt glass in a
continuous tank furnace to meet an exacting colour specification, it presents considerable problems.

There is an extensive technical literature on coloured
glasses and a valuable summary of this up to 1945 is to be

found in the book by Weyl (1951). A more recent account by Bamford (1977) is written primarily from the technological point of view, whilst Wong and Angell (1976) concentrate on spectroscopic aspects and the relationship between colour and glass structure.

In recent years, developments in transition metal ion chemistry have greatly added to the understanding of the structure and spectroscopy of compounds of these elements. A very readable introduction to the subject is available in a book by Orgel (1960). The results of this work have been applied with considerable success to give a satisfactory account of the absorption spectra of transition metal ion colours in glass. Progress has also been made in explaining, in terms of chemical thermodynamics, the factors that determine the equilibria between the various valence states. Review articles covering much of this work exist and these will be referred to later. The main purpose of this chapter is to provide an elementary summary at the same level as the rest of the book, with the hope of helping the reader to follow the more authoritative reviews with profit.

1. Effects of the Ligand Field

Table XXI lists the ions which are of the most interest in

TABLE XXI

Group 3 transition element ions and their colours

Ion	Number of d electrons	Ground state of the isolated ion	Colour in a soda-lime-silica glass
Ti^{3+}	1	^2D	Violet
V^{3+}	2	^3F	Green
Cr^{3+}	3	^4F	Green
Mn^{3+}	4	^5D	Purple
Mn^{2+}	5	^6S	Colourless
Fe^{3+}	5	^6S	Yellow-green
Fe^{2+}	6	^5D	Blue
Co^{2+}	7	^4F	Intense blue
Ni^{2+}	8	^3F	Grey brown
Cu^{2+}	9	^2D	Blue

making coloured glasses and the colours which they produce when present in a soda-lime-silica glass similar in composition to container or window glass. The table also gives the number of d electrons and the lowest energy state or ground state of the isolated ion. These symbols will be referred to again later, when it will be appropriate to give some indication of what they signify.

An isolated transition metal ion, free from the disturbing influence of nearby ions or other sources of electrical fields, may exist in a number of states, each characterised by an energy value. Such a series of energy levels, each labelled by a spectroscopic symbol, is shown in Fig. 108 for an ion containing

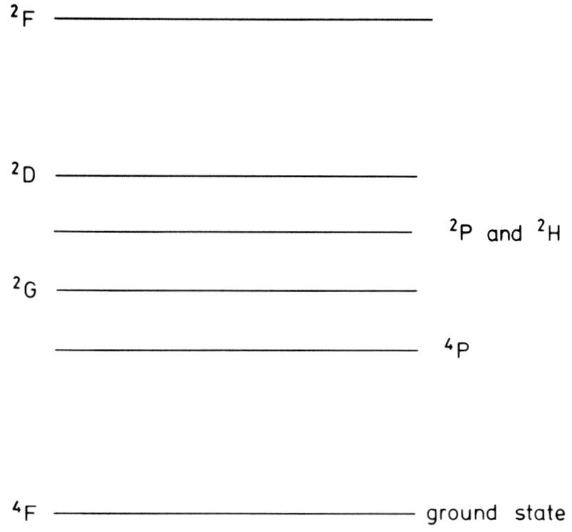

^{2}F ————————————————

^{2}D ——————————————

—————————————————— ^{2}P and ^{2}H

^{2}G ——————————————

—————————————————— ^{4}P

^{4}F —————————————— ground state

Fig. 108. Energy levels of an isolated ion containing three d electroncs.

three d electrons, e.g. the Cr^{3+} ion. One might expect such an ion to absorb radiation by excitation of the ion from its ground state to *any* one of the higher energy levels. However this is not so. Certain selection rules exist, derivable from quantum mechanical principles, which state that certain types of transition between energy levels cannot occur as a result of the absorption of radiation. One such, Laporte's rule, forbids transitions between initial and final states which involve the excitation of an electron from one d orbital to another. If this also applied when the transition ion is incorporated in a

glass, the ion would produce no colour. However, Laporte's rule
and other rules are relaxed to some extent when the ion finds
itself in an environment in which it experiences an electric
field which is not entirely symmetrical and which may be varying rapidly as a result of the thermal vibrations of the ion itself and those which surround it.

The selection rules still have some influence on the transitions of the ion when it is no longer isolated. Transitions
which are forbidden for the isolated ion are considerably less
probable than those which the rules allow. Consequently the absorption bands corresponding to the forbidden transitions are
relatively weak.

Another consequence of the incorporation of the ion into a
material is that most of the energy levels of the isolated ion
split up into a considerably greater number of levels. The
separation between these levels is greatly affected by the resultant field which the transition metal ion experiences and by
interactions between the ion and its immediate neighbours. Subject to the various selection rules, transitions between this
greater number of energy levels are possible. The energy difference between the ground state and the excited state, and
hence the wavelength at which the absorption occurs, depends
upon the immediate environment of the ion. In particular it
depends upon the symmetry of the electric field which the ion
experiences due to its immediate neighbours or ligands. The
type of spectrum obtained when the ion is surrounded by four
ligands each at a corner of a tetrahedron is greatly different
from that obtained when the ion is surrounded by six ligands at
each corner of an octahedron. The spectrum is also affected,
though to a lesser extent, by the nature of the ligands. For
octahedral co-ordination, the spectrum has a similar form re-
gardless of the nature of the ligand, but the absorption bands
occur at different wavelengths according to whether the ligands
are, for example, oxygen ions as in a glass or water molecules
as they would be in an aqueous solution.

The simplest example to consider is that of the Ti^{3+} ion
which has only one d electron. The free ion has a single level
denoted by the spectroscopic symbol ^2D. When the ion finds itself in an octahedral field, surrounded by six ligands, this

level splits into an upper and a lower level as indicated in
Fig. 109a.   This shows an increasing separation between the

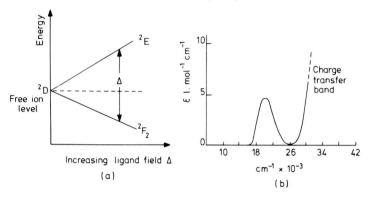

Fig. 109   a. Orgel diagram for the energy levels of a Ti^{3+}
ion in an octahedral ligand field. b. Absorption spectrum of
an aqueous solution containing $[Ti(H_2O)_6]^{3+}$ ions.

levels as the field due to ligands increases. Absorption of
radiation can occur involving a transition between the lower and
the upper levels. This gives rise to a single absorption band
at 20,000 cm^{-1} when the ion is surrounded by six water molecules
as it is in an aqueous solution (Fig. 109b). In a borosilicate
glass the spectrum has the same shape but the band is shifted to
about 17,500 cm^{-1}.

   For ions containing several d electrons, the energy level
diagram is more complicated. Fig. 110 is the diagram for Ni^{2+},
which contains 8 d electrons in octahedral co-ordination. One
should bear in mind that this and similar diagrams are a result
of theoretical calculations. Fortunately this very complicated
diagram does not result in an equally complicated spectrum.
Another selection rule, the multiplicity selection rule, considerably restricts the number of absorption bands which are
relatively strong in the spectrum.   The strong bands are those
involving transitions between states having the same value of
the spin multiplicity, this being given in the diagram by the
superscripts in the spectroscopic term symbol.   Thus the
rule allows transitions between the lowest energy state $^3A_2$,
which has a multiplicity of 3 and the higher level states
$^3F_2$ (F), $^3F_1$ (F) and $^3F_1$ (P). These are indicated by stars in
the figure. (It is unfortunate that different authors use
different symbols for the levels of the ions surrounded by

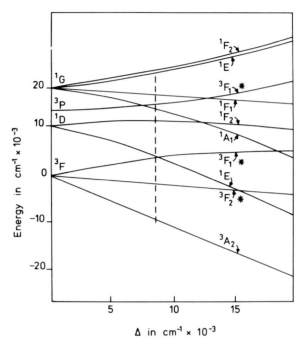

Fig. 110. Orgel diagram for the energy levels of the Ni^{2+} ion in an octahedral ligand field.

ligands. Even review articles are not internally consistent.)
The three absorption bands corresponding to the multiplicity-
allowed transitions of Ni^{2+} in octahedral co-ordination are
shown in Fig. 111. The bands occur at slightly different
frequencies according to whether the ion is in aqueous solution,
in a glass or in an oxide crystal, but clearly the spectra are
very similar in general shape. The bands occur at positions
corresponding to a Δ value of approximately 8,500 cm^{-1} as can
be established by reading off from Fig. 110 the vertical distances between the curves corresponding to the initial and final
energy levels at this value of Δ.

Figure 112 shows that the Ni^{2+} ion gives quite a different
kind of spectrum when it is in four-fold co-ordination. To
interpret this one has to use a different energy level diagram.
Generally speaking, the absorptions of an ion in four-fold co-
ordination are significantly stronger than in six-fold co-ordination. The ligand field is less symmetrical and the multiplicity rule is relaxed to a greater extent.

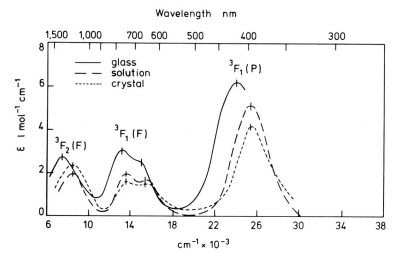

Fig. 111. Absorption spectra of various media containing octahedrally co-ordinated Ni^{2+} ions (Bates, 1962).

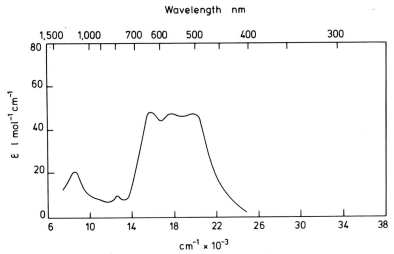

Fig. 112. The absorption spectrum of Ni^{2+} in a high alkali borosilicate glass (Bates, 1962).

A further important consequence of introducing a transition metal ion into a material such as a glass or an aqueous solution is that the energy of the ion in its ground state is changed relative to that of the free ion. The magnitude of this change depends on the nature of the ion and whether it finds itself in four-fold or six-fold co-ordination. For example, the theory shows that Ni^{2+} in octahedral co-ordination has a lower energy than in four-fold co-ordination; whereas for Co^{2+} the energy difference between the two forms of co-ordination is relatively small.

For a considerably more detailed account of the interpretation of the spectra due to transition metal ions in glasses the reader is referred to Bates (1962) and Wong and Angell (1976). Even before the development of the current interpretation of transition metal ion spectra had been developed, glass scientists had made considerable use of the variation of the absorption spectra with glass composition to deduce what changes in glass structure might be occurring as a result of a composition change. Although such deductions may still be debatable in particular instances, there now exists a much sounder foundation for interpretations of this kind.

2. Effect of Glass Composition on the Colours due to Co^{2+} and Ni^{2+} in Oxide Glasses

It is worth looking briefly at the effects of changes of glass composition on the absorption spectra due to these two ions. The reason for choosing these in particular is that in almost all silicate, borate, and phosphate compositions melted under normal conditions only the divalent ions are present. One does not have the additional complication of a mixture of valence states.

As the glass composition is made more basic by increasing the alkali content, the colour of borate and silicate glasses containing Co^{2+} changes from pink to an intense blue. This is due to a change in the oxygen co-ordination number of the cobalt ion from 6 (octahedral symmetry) to 4 (tetrahedral symmetry). Figure 113 due to Paul and Douglas (1968) shows this effect for glasses in the $Na_2O-B_2O_3$ system. Figure 114 shows how the area under the absorption curve between the wavelengths 430 and 710 nm increases with increasing R_2O content for three alkali borate systems. Paul and Douglas suggest that the inflection at 20 mol% R_2O indicates that the change to tetrahedral co-ordination of the Co^{2+} ion begins at this composition. The curves also indicate that, above the inflection, the K_2O glasses contain the highest proportion of Co^{2+} in this co-ordination and the Li_2O glasses the least.

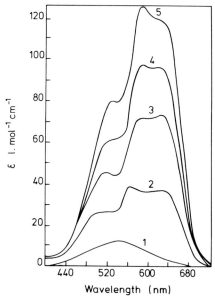

Fig. 113- The absorption spectra of Na_2O-B_2O_3 glasses containing Co^{2+}. Na_2O contents in mol%: 1, 13%; 2, 22.5%;
3, 26.5%; 4, 30.2%; 5, 33.0%.

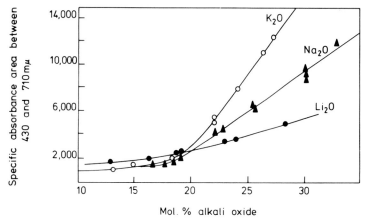

Fig. 114. The effect of alkali content in alkali borate
glasses on the strength of the absorption due to Co^{2+} in
tetrahedral co-ordination. (Paul and Douglas, 1968).

All the alkali silicate glasses containing CoO have a similar absorption spectra to the high alkali borates. They are
deep blue in colour and presumably contain only Co^{2+} in four-
fold co-ordination.   This has been attributed to the fact that

the initial addition of alkali to a silicate melt produces non-
bridging oxygen ions whereas this does not happen in the borate
system.   The availability of non-bridging oxygens has been regarded as favouring the increase in the co-ordination number of
the Co^{2+} ion.

Changes in glass composition have a similar effect on the
symmetry of Ni^{2+} ions and hence on the colour of the glass.
Increasing the alkali content results in a change from octahedral to tetrahedral symmetry, although it has been suggested that
there may be an intermediate region in which the Ni^{2+} ions are
surrounded by four oxygen ions at the corners of a square.

There is an extensive literature on the interpretation of
the absorption spectra of glasses containing Co^{2+} and Ni^{2+} in
terms of structural changes occurring in the glass (Paul, 1971).
However, changes in the spectrum may not always be a *direct* result of a change in the proportion of non-bridging oxygens.
Paul (1974) has pointed out that, at least in alkali phosphate
glasses, the proportion of Ni^{2+} in tetrahedral co-ordination in
a particular glass increases with increasing temperature.   He
suggests on this basis that for glasses in the $Na_2O-CaO-P_2O_5$
system, changes in the symmetry of the Ni^{2+} ions for glasses of
differing CaO content may be due to the effect of CaO on the
transformation temperature of the glass, i.e. the equilibrium
between tetrahedral and octahedral co-ordination is frozen in at
a higher temperature as the CaO content of the glass is increased.

3. Factors Affecting the Redox Equilibria of Transition
    Metal Ions in Glass

Most of the transition metals are multivalent, and when an
oxide of one of these metals is dissolved in glass an equilibrium
is established between two types of ion representing different
valence states of the metal.   The colour of the glass depends
on the relative proportions of these ions.   The equilibrium
ratio is affected by a number of factors: the partial pressure
of oxygen in the atmosphere of the furnace in which the glass
has been melted, the melting temperature, the composition of the
glass and the presence of other multivalent ions in the glass.
An equilibrium ratio is established only if the melting temperature is high enough and the melting time is sufficiently long.
Even at melting temperatures of the order of $1400°C$ it may take

many hours for equilibrium to be reached in a silicate glass.
When melting such glasses continuously on a tonnage scale in a
large tank furnace it is doubtful if equilibrium with the furnace
atmosphere is established in more than a small fraction of the
glass melted.

a. Effects of Furnace Atmosphere and Melting Temperature

This and the following section will be concerned with the
simple situation in which only one multivalent element is involved. A number of studies have been made of the effects of
the oxygen partial pressure in the laboratory furnace atmosphere
and the melting temperature on the equilibrium between the valence states of a number of transition elements dissolved in
glasses of simple composition.   Figure 115 shows the results of

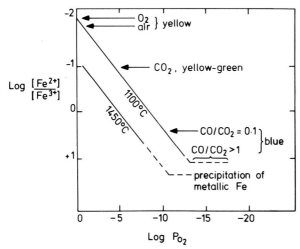

Fig. 115. The effect of the partial pressure of oxygen on
the $[Fe^{2+}]/[Fe^{3+}]$ ratio in a $Na_2O-2SiO_2$ glass (Johnston, 1964).

Johnston (1964) for the equilibrium between the ferrous and
ferric ion in a glass of composition $Na_2O.2SiO_2$ containing 2 wt%
iron oxide.   The equilibrium between the ions and the furnace
atmosphere may be expressed by the equation
$$4Fe^{2+} + O_2 \rightleftharpoons 4Fe^{3+} + 2O^{2-} .$$
The oxidation of the ferrous ion is accompanied by the transfer
of oxygen to the glass from the furnace atmosphere, the oxygen
atoms acquiring electrons from the ferrous ions, thus changing
into oxygen anions which are incorporated into the glass struc-

ture. The low rate of diffusion of the oxygen ions, particularly in silicate melts, is the reason why the rate of attain-
ment of equilibrium with the atmosphere is so slow. Once equilibrium has been attained, the glass is rapidly cooled to room
temperature. The high temperature equilibrium is thereby frozen in and the relative proportions of the two ions can be conveniently determined at room temperature by chemical and physical methods. The equilibrium constant K of the Fe^{2+}/Fe^{3+} reaction is given by

$$K = ([Fe^{3+}]^4 \cdot [O^{2-}]^2)/([Fe^{2+}]^4 \cdot p_{O_2}) \qquad (94)$$

where $[Fe^{3+}]$, $[Fe^{2+}]$ and $[O^{2-}]$ represent the activities of these
ions in the glass and p_{O_2} is the partial pressure of oxygen in
the furnace atmosphere. Since the concentrations of the iron
ions in the glass are low, it is reasonable to assume that their
activities are proportional to their concentrations. One can
also assume that the oxygen ion activity in a given glass and
at a given temperature is constant and is not significantly affected by any change in the balance of the reaction. One would
therefore expect the following equation to hold

$$\log((Fe^{2+})/(Fe^{3+})) = \log([Fe^{2+}]/[Fe^{3+}]) = \log p_{O_2}/4 \qquad (95)$$

(Fe^{2+}) and (Fe^{3+}) are the concentrations of the respective ions
in the glass. Johnston's results fit this equation very well.

Increasing the melting temperature increases the proportion
of Fe^{2+}. Thus for the glass which has been brought into equilibrium with air (p_{O_2} = 0.2) at 1100°C only 1.5% of the iron is
present as Fe^{2+}. At 1450°C this increases to 9%. Because the
iron is predominantly ferric, the colour of both glasses at room
temperature is yellow. By melting in furnace atmospheres which
are mixtures of CO and CO_2, very low oxygen partial pressures
are readily obtained and the oxygen partial pressure can be calculated from the composition of the gas mixture. The results
show the expected increase in the $(Fe^{2+})/(Fe^{3+})$ ratio as the
CO/CO_2 ratio is increased. The glasses containing high percentages of ferrous iron are blue in colour.

A number of similar studies have been made of the redox
equilibria of other transition elements. Thus Johnston (1965)
investigated the equilibria Ti^{3+}/Ti^{4+}, Ce^{3+}/Ce^{4+}, Mn^{2+}/Mn^{3+},
Sb^{3+}/Sb^{5+} and Sn^{3+}/Sn^{4+} and Banerjee and Paul (1974) in a study

of copper ruby glasses investigated the equilibria Cu^+/Cu^{2+} and
Cu^0/Cu^+ in borate glasses. In all these investigations, increasing the melting temperature and decreasing the partial pressure of oxygen in the atmosphere has the same effect of moving
the balance of the equilibrium towards the lower valence state.

As a final point, it may be noted that investigations of
a number of redox equilibria have shown that, at very low concentrations of the transition element, the equilibrium shifts
significantly to the lower valence state (see, for example,
Paul and Douglas, 1965). The reason for this is not known.

b. The Effect of Glass Composition

The general effect of making the glass composition more
basic, i.e. of increasing the R_2O content in binary borate,sil-
icate and phosphate systems is to increase the proportion of the
transition element in the higher valence state. The melt becomes more oxidizing. Figure 116 shows results for the

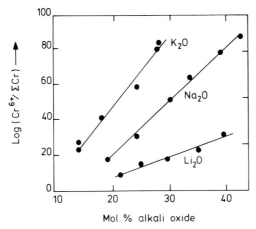

Fig. 116. The effect of glass composition on the ratio
$(Cr^{6+})/(Cr^{6+} + Cr^{3+})$ in alkali silicate glasses. (Nath
and Douglas, 1965).

Cr^{3+}/Cr^{6+} equilibrium in alkali silicate glasses (Nath and
Douglas, 1965). At a given molecular percentage of R_2O, the
K_2O melt is most oxidizing and the Li_2O melt the least.

There is a considerable amount of discussion in the literature on how best the interpret these results. Following the
procedure of the previous section, one would write the equation
for the Cr^{3+}/Cr^{6+} redox reaction as follows

$$Cr^{3+} + 3/4.O_2 = Cr^{6+} + 3/2.O^{2-}$$

Nath and Douglas, in discussing their results, point out that this leads to an obvious difficulty. Increasing the alkali content of the glass will increase the oxygen ion activity and this should displace the equilibrium to the left - the opposite to what is observed. They suggest that this must imply that the equilibrium constant of the reaction must change with glass composition. In a recent paper Douglas (1974) amplifies this hypothesis by pointing out that the activity of an ion in solution is not in general equal to its concentration. The two are related by an activity coefficient γ. For example $[Cr^{6+}] = \gamma(Cr^{6+})$. He gives reasons why the activity coefficient for Cr^{6+} may decrease with increasing alkali content of the glass. One has also to consider, of course, the activity coefficient of the Cr^{3+} ion and how this should vary with composition. It is interesting to compare this approach with that of Budd (1966) who proposed that it would be preferable to consider the Cr^{3+}/Cr^{6+} equilibrium as involving the chromate ion CrO_4^{2+} in which case the redox reaction might be written

$$CrO_4{}^{2-} = Cr^{3+} + 2.5\ O^{2-} + 0.75\ O_2\ .$$

This equation makes it easier to see why an increase in alkali content should increase the proportion of chromium in the hexavalent state. This problem is discussed at some length by Wong and Angell (1976) who appear to incline to an interpretation which is similar to Budd's hypothesis.

This is certainly an area where it is appropriate to use the time-honoured phrase "further work is required" as an euphemism for "we don't know". No doubt many uncertainties will be resolved if a convenient and reliable method can be developed for measuring oxygen activities in glass melts, or if some quite different interpretation of the results is found to be more successful (see Duffy et al., 1978; Wagner, 1975).

4. Interaction between Redox Ions in Glass

The previous section has shown that there is some uncertainty as to how to interpret some of the experimental results obtained in the study of redox equilibria in glass melts. Not surprisingly, similar but even greater problems arise when one is dealing with a situation in which the glass contains more than one multivalent element. One has to contend with such situations in practice. Some commercial coloured glasses do contain more than one multivalent transitional metal and the common practice of decolourising glass involves an interaction between the As^{3+}/As^{5+} or Ce^{3+}/Ce^{4+} redox couples and the Fe^{2+}/Fe^{3+} couple.

When a glass contains two redox couples, the equilibrium between the valence states is not influenced by the partial pressure of oxygen in the furnace atmosphere. Thus in a glass containing, for example, small percentages of the oxides of cerium and chromium, the reaction between the ions might be described by the equation

$$3Ce^{3+} + Cr^{6+} = Cr^{3+} + 3Ce^{4+}$$

The reduction of Cr^{6+} to Cr^{3+} occurs simply by a transfer of electrons from the Ce^{3+} ions. This can occur rapidly; no interaction with the furnace atmosphere is required. There is therefore considerable doubt whether the proportions of the various ions determined after cooling the melt to room temperature is the same as that which existed at the melting temperature.

A number of studies have been made of the equilibria between pairs of redox oxides in glass melts. Among the most recent are those of Paul and Douglas (1966) and Lahiri et al. (1974). The results have been discussed in terms of the thermodynamics of the postulated reactions by Paul and Douglas (1966) and Douglas (1974). The commonly used starting point for these discussions is an Ellingham diagram which shows the variation with temperature of the change in free energy when an oxide reacts with one mole of oxygen to form a second oxide in which the metal has a higher valency. Such a diagram is shown in Fig. 117. It is important to remember that this and similar diagrams refer to the pure oxides and not to these oxides when dissolved in a glass.

To illustrate the use of an Ellingham diagram, consider the following reactions:

Reaction 1 $As_2O_3 + O_2 \rightarrow As_2O_5$

Reaction 2 $4FeO + O_2 \rightarrow 2Fe_2O_3$

At $1600°K$ the free energy changes for these reactions are respectively $\Delta G_1 = +20$ K Cal and $\Delta G_2 = -50$ K Cal. If one now considers the reaction

212

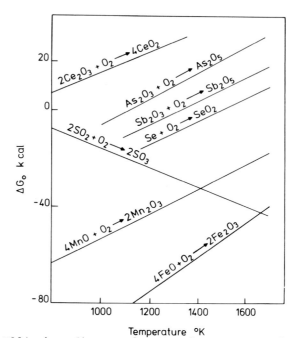

Fig. 117. Ellingham diagram for the free energy of reaction
per mole of oxygen of some redox systems of interest in glass
making.

$$4FeO + As_2O_5 \rightarrow 2Fe_2O_3 + As_2O_3$$

the free energy change will be $\Delta G = \Delta G_1 - \Delta G_2 = -70$ K Cal.
Consequently the reaction will go to the right and As_2O_5 will
oxidize ferrous oxide to ferric oxide. Similarly one can show
that any redox pair lying near the top of the diagram will oxidize those lying below it. The Ce_2O_3/CeO_2 and Cr_2O_3/CrO_3 (not
shown) pairs, which are close together near the top of the diagram, are the most oxidizing and should, for example, completely
oxidize FeO to Fe_2O_3.

When these oxides are dissolved in glass it is not to be
expected that the behaviour will be exactly as predicted by the
Ellingham diagram. However it is found that for the reactions

$$Ce^{4+} + Fe^{2+} \rightleftharpoons Fe^{3+} + Ce^{3+}$$
$$Cr^{6+} + 3Fe^{2+} \rightleftharpoons 3Fe^{3+} + Cr^{3+}$$
$$As^{5+} + 2Fe^{2+} \rightleftharpoons 2Fe^{3+} + As^{3+}$$

the equilibrium is predominantly to the right as predicted, unless there is insufficient cerium, chromium or arsenic oxide
present for complete oxidation of the iron to be possible.

When dealing with redox pairs which lie closer together, the Ellingham diagram is of little value for predicting what will happen when they are brought together in a glass. Thus the diagram predicts that CeO_2 will oxidize Cr_2O_3 to CrO_3, the free energy change being -10 K Cal for the reaction

$$4CeO_2 + 2/3.Cr_2O_3 = 4/3.CrO_3 + 2Ce_2O_3.$$

However, although the equilibrium is to the right in a sodium aluminoborate glass it is to the left in a soda-lime-silica glass.

To be able to understand the interaction of redox pairs in glasses of various compositions, one needs information about the free energy changes that occur when the oxides are dissolved in the glass in question. Expressed in another way, one needs to know the activities of each ion. Douglas (1974) has discussed some of the available experimental results in terms of the probable changes of chemical potential which occur when oxides are dissolved in glass and he develops a modified Ellingham diagram which gives a convenient pictorial representation of the changes that occur and of the changes in the equilibria to be expected as the glass is cooled. These considerations are of considerable interest and may be correct, but much more experimental work is clearly needed before a completely satisfactory account can be given of the behaviour of transition metal ions in glass.

D. The Absorptivity of Oxide Glasses in the
   Ultra-Violet

The absorption of ultra-violet radiation by oxide glasses has been referred to in the previous chapter during the discussion of the dispersion properties.

As one moves into the UV from the visible, the absorption coefficient of oxide glasses begins to rise rapidly and most of these materials are effectively opaque at wavelengths shorter than 200 nm in thicknesses of a millimetre or so. The rapid rise of absorption coefficient is referred to as the "UV cut off" and many studies have been made of the effect of composition on the wavelength at which it occurs. The results of work of this kind are, of course, valuable in the development of UV transmitting glasses for such applications as bactericidal lamps. The interpretation of the results is complicated by the fact

that the increase in absorption in the UV is determined partly
by the major components of the glass composition and partly by
impurities, in particular the oxides of iron, chromium and titanium which may be present at levels of only a few parts per
million. The ions Fe^{3+}, Cr^{3+} and Ti^{3+} absorb very strongly in
the ultra-violet. Thus the molar absorptivity of Fe^{3+} in a
silicate glass is approximately 3,000 l mole^{-1} cm^{-1} and is centred at about 210 nm. This is about one hundred times greater
than the absorptivities characteristic of the electron transitions referred to in the previous section. The electron transitions responsible for these strong absorptions are called
charge transfer transitions because they are believed to involve
excitations in which an electron leaves an orbital mainly located
on one ion, e.g. an Fe^{3+} ion to occupy an orbital mainly located
on a nearby ion, e.g. one of the nearest-neighbour oxygen ions
(or vice versa).

Since, as we have seen, a change in glass composition af-
fects redox equilibria such as Fe^{2+}/Fe^{3+} and Cr^{3+}/Cr^{6+}, it is
difficult to decide whether a shift in the UV cut off is associated with electron transitions in the ions which form the major part of the glass structure or is merely due to changes in
the valence state of impurity ions. Only relatively recently
has work been done on glasses made from sufficiently pure raw
materials to be reasonably certain that composition effects on
the UV cut off can be attributed to transitions involving the
major constituents of the glass (Sigel, 1973). As stated in
the previous chapter, the main effect of increasing the alkali
content of a silicate or borate glass is to move the UV cut off
to longer wavelengths. It appears to be generally accepted
that the first strong absorption encountered on moving to shorter wavelengths in the UV involves the excitation of an electron
from valence levels on oxide ions to higher levels localized on
the same ions. Also, for non-bridging oxygens the excitation
energy is less than for bridging oxygens; consequently the absorption occurs at a longer wavelength.

Since it is clearly impossible to study the UV absorption
spectra directly because of the high absorptivity of oxide
glasses in this region (10^4-10^6 cm^{-1}), indirect methods must be
used. A certain amount of information is available on the vari-

ation of the reflectivity of oxide glasses with wavelength in
the UV.   From the results of experiments of this kind it is
possible to calculate the variation of absorptivity with wavelength.   For oxide glasses, the wavelengths of the absorption
maxima are fairly close to the wavelengths of the reflectivity
maxima.   Figure 118 shows the results of Phillipp(1966) for

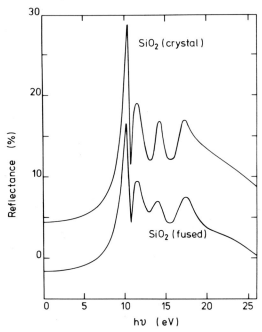

Fig. 118.   Reflection spectra of crystalline and vitreous
silica.

α-quartz and vitreous silica.   Note that the reflection maxima
occur at practically the same energy values (or wavelengths) for
both materials.   The absorption maxima occur at wavelengths
(calculated from these results) of 10.2, 11.7, 14.3 and 17.2 eV.
The wavelength corresponding to the lowest energy peak is 121.5
nm.   This is a wavelength significantly less than that of the
UV cut off for high purity vitreous silica which is about 160
nm.   Sigel (1973) has measured the reflection spectra of various two- and three-component alkali silicate glasses and has
observed additional maxima at 8.5 eV (146 nm) and 9.3 eV (133
nm).   The position of these maxima is practically independent
of the nature of the alkali.

In recent years solid state physicists have become increasingly interested in amorphous materials. A factor which has stimulated this interest is the discovery of semi-conducting glasses (Chapter VIII) and the realization that these glasses may have important technological applications. Another important factor has been the purely intellectual challenge of extending to materials having a non-periodic structure the theoretical methods which have been so successful in interpreting the properties of crystalline materials. Both the electrical and the optical properties of materials are dependent on the energy levels in a material and the distribution of the electrons between these levels. It is not surprising therefore that the UV absorption properties of simple glasses, especially silica, have been subjected to intensive experimental and theoretical study in recent years. A good impression of the depth and intensity of this work is given in a review article by Griscom (1977).

E. Colloidal Colours

The optical absorptions discussed so far are due to electron transitions in materials that are electrical insulators. Such materials are transparent in the visible because only UV photons have sufficient energy to excite the valence electrons whilst the frequency of the lattice vibrations which determine the infra-red absorption spectra (section F) fall outside the visible frequency range.

Metals, on the other hand, are opaque because they contain many electrons in the partly filled conduction band. These can be excited to higher energy levels within the band by interaction with visible light photons. Most semi-conductors also absorb strongly in the visible.

Some metals and semi-conductors are slightly soluble in oxide glasses. By suitable heat treatment at temperatures below the melting point they can be made to precipitate out from solution to give a large number of particles of colloidal size. The glass is coloured, the colour depending on the optical properties of the precipitated metal or semi-conductor and on the size and concentration of the particles.

The materials most commonly used in producing colloidal colours in glass are listed in Table XXII.

TABLE XXII

Colours produced in glass by colloidal particles of metals
and semi-conductors

Material	Colour
Au	Ruby
Ag	Yellow or amber
Cu Cu_2O	Ruby
Se	Pink
CdS	Yellow
CdSe	Ruby

In making these materials the composition of the glass and
the melting schedule are critical.   For the gold, silver and
copper colours, reducing agents must be added to the batch and
for the glasses containing Se, CdS and CdSe problems arise which
are due partly to the ease with which these materials are oxidized and partly to their loss by volatilization during melting.

The copper and silver colours may also be produced by surface treatment processes in which ions of the colouring material
are diffused into the glass at relatively low temperatures.   The
staining of glass by silver compounds to produce yellow or amber
colours has been known for centuries and has been a widely used
method for making coloured church windows since the early Middle
Ages.   A paste consisting of an inert carrier such as clay or
ochre mixed with the silver compound, usually the chloride or
the sulphide, is applied to the glass surface.   The glass is
then heated to a temperature a little above its annealing temperature.   A base exchange process takes place between silver
ions in the paste and alkali ions in the glass.   The silver
ions are reduced in the glass to atoms of metallic silver, which
coalesce to form particles of colloidal size.   Red copper stains
can be produced in a similar way.   Several workers have studied
the mechanism of the base exchange reaction (e.g. Meistring et
al., 1976).

In another method of producing a surface colloidal colour,
the metal ions are introduced into the glass by electrolysis,
an equal number of alkali ions being removed through the oppo-

site surface. This principle is used in the Pilkington "Spectrafloat" process, one of the electrodes being the bath of molten tin on which the moving ribbon of glass is formed (Fig. 119).

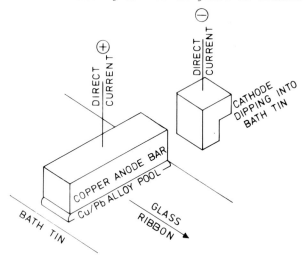

Fig. 119. The "Spectrafloat" process.

The colourant metal comes from the metallic pool trapped between
the anode and the upper surface of the glass. This pool consists of a low melting point alloy, one constituent of which is
the colourant metal. Electrical charge balance is maintained
by the transfer of alkali ions at the lower surface of the ribbon from the glass to the tin bath (Yates, 1974). The presence
of a reducing atmosphere over the bath ensures the reduction of
the metallic ion diffused into the upper surface of the glass.

The first major application of the process was to produce
a bronze coloured material which has been widely used for solar
control windows. The colour is due to particles of a copper-
lead alloy. More recently, the technique has been developed
to produce a variety of decorative effects. By periodically
interrupting the current flow coloured stripes are produced,
the shape of which can be determined by suitable design of the
upper electrode. Also by using a number of electrodes, each
working with an alloy of different composition, multi-coloured
stripe patterns can be produced.

Most of the scientific study of the colloidal colours has
been concerned with relating the visual absorption spectrum of

the glass to the optical properties of the colloid material and
with studies of the nucleation and growth of the particles. The
absorption spectrum can be calculated by applying the Mie theory
for the absorption and scattering of light by colloidal particles.
Provided that the particles are small compared with the wavelength of light and that they are so far apart that they scatter
independently, the absorption coefficient is given by

$$\gamma_\lambda = A_\lambda NV \qquad (96)$$

where N is the number of particles per unit volume of glass, V
is the particle volume and A_λ is given by

$$A_\lambda = (36\pi n_0{}^3/\lambda)(nk)/((n^2-k^2+2n_0{}^2)^2 + 4n^2k^2) \qquad (97)$$

n_0 is the refractive index of the glass and n and k are the optical constants of the particle material.

Studies by Maurer (1958) and Doremus (1964, 1966) on glasses
coloured by silver and gold have shown that good agreement is obtained between the measured spectra and the spectra calculated
using the Mie equations.   Figure 120 shows the comparison for

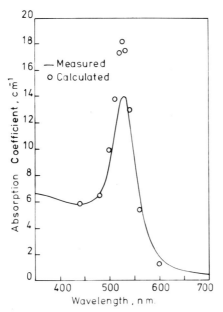

Fig. 120. Absorption spectrum of gold particles about
120 A.U. in diameter in glass (Doremus, 1964).

a glass coloured by gold particles which are approximately 120

A.U. in diameter.   A similar comparison for a glass coloured
by silver particles is shown in Fig. 121.   Bearing in mind the

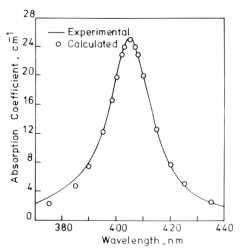

Fig. 121. Absorption spectrum of silver particles about
100 A.U. in diameter in glass (Doremus, 1966).

difficulty of making reliable measurements of the optical properties of metals, the agreement is very good.   Comparisons of
measured and calculated absorption spectra for copper stains
have been made by the present author (Rawson, 1965) and for
"Spectrafloat" by Bamford (1976).   In both these studies it
is assumed that the ruby colour is due to particles of metallic
copper.   It would be surprising if this were not so since the
materials studied had been exposed to hydrogen-containing atmospheres.

However there is strong evidence suggesting that the colours of some copper ruby glasses are due to colloidal particles
of cuprous oxide.   This was in fact suggested by Atma Ram and
Prasad (1962).   More recently Banerjee and Paul (1974) have
carried out a detailed thermodynamic study which supports this
view, at least for a copper ruby produced in sodium borate
glasses.   They discount the evidence based on agreement between the measured and calculated absorption spectra on the
grounds that the n and k measurements which have been made on
copper are unreliable.   They consider that the films on which
these measurements were made would have been heavily contaminated with cuprous oxide.

There is also a difference of opinion concerning the cause of the colour of selenium pink glasses. Both Weyl (1951) and Bamford (1977) suggest it is due to a molecular solution of elemental Se in the glass. However Paul (1975) has shown that selenium pink glasses which he produced contained colloidal particles and that the absorption spectrum could be predicted using the Mie equations.

The yellow CdS and the ruby CdSe colours are almost certainly due to colloidal particles. These glasses are of considerable technological interest because of the very sharp transmission cut-off in the blue part of the spectrum (Fig. 122) and

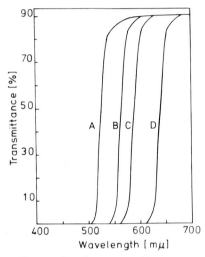

Fig. 122. Curves of spectral transmittance of glasses containing CdS alone (A) and with increasing amounts of CdSe (B,C,D). (Bausch and Lomb Inc.).

the fact that the wavelength of this cut-off can be controlled over a fairly wide range by using mixtures of CdS and CdSe in the glass composition. The absorption spectrum can also be controlled by varying the heat treatment (Izumitani and Matsuura, 1967).

Colloid colours may be spoiled by too lengthy a heat treatment causing the particles to grow to a size comparable with the wavelength of light. The scattering of light becomes noticeable to the eye and the colour looks turbid or muddy.

F. Absorption in the Infra-red

The absorption spectra of glasses in the infra-red is largely determined by interactions between the material and radiation which excite atomic vibrations. These may involve oscil-
latory changes in interatomic distances (bond stretching) or
in bond angles (bond bending). For oxide glasses at least,
there are only one or two instances of IR absorption bands due
to the excitation of electrons. The most important of these
technologically is that due to ferrous iron at about 1.1 μm.

For a simple diatomic molecule, absorption of radiation
may occur at the natural vibration frequency of the molecule.
This can be calculated using classical mechanics. If the force
constant of the bond is f and the masses of the atoms in the
molecules are m_1 and m_2, then the vibrational frequency is
given by

$$\nu = 1/2\pi . (f/\mu)^{\frac{1}{2}} \tag{98}$$

where

$$\mu = m_1 m_2 / (m_1 + m_2)$$

Absorption of electro-magnetic radiation occurs if the molecule
has a dipole moment. Thus homonuclear molecules such as O_2
and N_2 have no vibrational infra-red spectrum. Although this
simple equation cannot be applied to materials as complicated
as inorganic glasses, nevertheless it is a general rule that
glasses having compositions based on heavy atoms bonded together
by relatively weak forces have their strong infra-red absorptions at longer wavelengths than glasses consisting of light
atoms bonded together by strong forces. In particular the
chalcogenide glasses absorb at longer wavelengths than the
silicate glasses.

For molecular substances the infra-red absorption spectrum
can give detailed information about the interatomic bonding
forces and the configuration of the molecule. The interpretation of infra-red spectra has proved to be a valuable tool in
the study of the structure of even quite complex organic molecules. The measurement of infra-red spectra is also widely
used as a routine method for the analysis of mixtures of organic
materials.

It is a more difficult problem to predict theoretically
the modes of vibration and the frequencies of atomic vibration
of solids which have an infinite three-dimensional structure,
unless the structure is a simple crystalline one. Only relatively recently have numerical calculations been carried out on
the vibrational spectra of materials having non-periodic structures. Consequently much of the work which has been done in
an attempt to interpret the infra-red spectra of oxide glasses
has simply involved comparisons between the spectra of the
glasses with those of crystalline compounds of known structure.
The more recent work which exploits the ability to calculate
the structural vibrations is at present limited in scope to
glasses of simple composition, e.g. SiO_2, GeO_2 and BeF_2.

A review article by Parke (1974) deals with both the scientific and the technological aspects of the infra-red absorption
spectra of glasses whilst a chapter in the book by Wong and Angell (1976) gives greater prominence to the more theoretical
work. When looking at the literature in this field one sees,
as in the study of UV absorption, two types of work. One is
concerned with the development of IR transmitting or absorbing
glasses, when one is dealing with glass thicknesses of a few
millimetres (or very much more as in fibre optics), whilst the
other is concerned simply with identifying the strength and positions of the main infra-red absorption bands. These are very
strong, with absorptivities of the order 10^4 cm^{-1} and are normally studied either by measuring the reflection spectra or, if
transmission is used, by greatly diluting the glass by pressing
the glass powder into a disc with potassium bromide. The wavelength at which a practically useful transmission can be obtained
is not directly determined by the wavelengths at which the main
absorptions occur but rather by overtone vibrations which occur
at simple multiples of the frequencies of the principal vibrations. Thus the first strong absorption of silica glass occurs
at about 10 μm, but practically useful transmission is limited
to about 4 μm by the presence of the overtone vibration at 5 μm.

A good, relatively simple example of the use of infra-red
absorption spectrum to infer the structure of an oxide glass is
shown in Fig. 123 for the crystalline and vitreous forms of
GeO_2. In hexagonal GeO_2, X-ray structural studies have shown

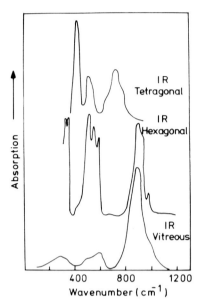

Fig. 123. IR spectra of tetragonal, hexagonal and vitreous
GeO$_2$. (Wong and Angell, 1976).

that the germanium atoms are surrounded by four oxygens at the
centre of an irregular tetrahedron.   In the tetragonal form,
the germanium atoms are surrounded by six oxygens.   The infra-
red spectrum of the glass, with its strongest absorption at 930
cm^{-1} suggests that the glass structure is more similar to that
of the hexagonal crystalline form, a suggestion which has been
supported by X-ray studies of the glass structure.

The absorption spectrum of silica shows a strong band at
1098 cm^{-1} and somewhat weaker bands at 800 cm^{-1} and 465 cm^{-1}.
The numerical studies of the vibration of the vitreous silica
structure by Bell and Dean (reviewed by Bell, 1972) suggest that
the band at 1098 cm^{-1} is a bond stretching vibration, the other
two being due to bond bending vibrations.

Considering the results of Hanna and Su (1964) for Na$_2$O-
SiO$_2$ glasses shown in Fig. 124, it appears that the progressive
introduction of Na$_2$O does not result in a marked change in the
infra-red spectrum.   As the alkali content is increased, the
absorption at 1100 cm^{-1} decreases in intensity and moves to lower frequencies and beyond 30 mol% Na$_2$O, this band splits with
one maximum at 1050 cm^{-1} and the other at 960-920 cm^{-1}.   It has

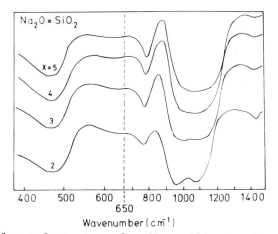

Fig. 124.   Infra-red spectra of sodium silicate glasses.

been suggested that the latter is due to non-bridging oxygens.
Absorptions due to vibrations of the alkali ions occur at wavelengths in the far infra-red (100-300 cm^{-1}).

A particularly extensive study of the absorption spectra
of alkali borate glasses has been carried out by Krogh-Moe (1965).
This is of particular interest for its use of comparisons with
the spectra of crystalline borates to infer the structures which
may be present in the glasses.

As pointed out earlier in this section, the chalcogenide
glasses transmit to much longer wavelengths in the infra-red than
do silicate glasses and for this reason they have been widely
studied, particularly for applications in heat-seeking missiles.
Transmission curves of a number of these glasses are shown in
Fig. 125.   Their manufacture involves taking great care to minimize oxide contamination.   Even a few parts per million of oxygen is sufficient to cause strong absorption bands.   Reviews
on the preparation and properties of these glasses have been
published by Savage and Nielson (1965), Hilton (1966) and Savage
et al. (1977).

If a glass is required which transmits to a slightly longer
wavelength than the silicate glasses, i.e. to 5-6 μm, materials
based on glasses in the CaO-Al_2O_3 system may be satisfactory.
The two-component glasses devitrify too easily to be useful and
although the addition of as little as 5 wt.% of SiO_2 greatly reduces this tendency, it also has a marked effect on the I.R.

226

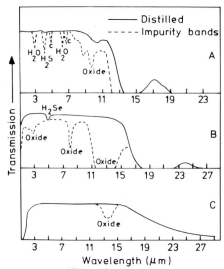

Fig. 125. Transmission spectra of (a) an arsenic sulphide glass, (b) a germanium-arsenic-selenide glass and (c) a germanium-arsenic-telluride glass. (Savage and Nielsen, 1965).

transmission. Consequently a considerable amount of development work has been done, again primarily for military applications, to develop stable but silica-free aluminate glasses (Hafner et al., 1958). This has included the development of vacuum melting techniques to reduce the OH content of the glasses.

Transmission of all oxide glasses in the near infra-red is affected by the presence of water, chemically combined in the glass as OH anions bonded to the silicon ions. In SiO_2 glass there is a single fairly narrow band at 2.85 μm whilst in Na_2O-CaO-SiO_2 glasses there are additional bands at 2.8 and 3.5 μm (Fig. 126). The intensities and positions of these bands which have been studied in some detail by Scholze (1959) vary with glass composition.

The other important constituent of silicate glasses affecting the transmission in the near infra-red is, as has already been mentioned, iron oxide when the iron is in the ferrous condition. Thus the IR transmission in this region is markedly affected by the total iron oxide content of the glass and by the proportion of the iron in the reduced state. The latter

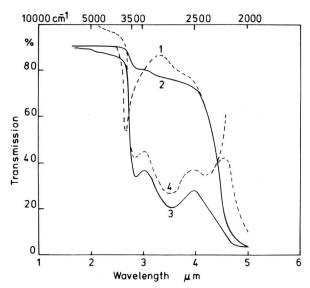

Fig. 126. Infra-red transmission spectra of some silicate glasses (1 mm thickness). 1: silica glass; 2: soda-lime-silica glass of low OH content; 3: the same glass as 2 but with a high OH content; 4: the difference spectrum between curves 2 and 3. (Scholze, 1977).

is of course affected by the melting conditions and by the presence of other redox pairs in the glass.

Since the heat transmission properties of the glass in the melting furnace are greatly affected by the absorption properties of glass in the near infra-red, it is of considerable importance for control of the melting process to ensure in particular that the Fe^{2+} and OH^- contents of the glass are controlled (see Gardon, 1961).

G. Decolourizing of Commercial Glasses

A considerable amount of effort is expended by glass manufacturers, particularly those in the container industry, in controlling the tint of their glass. Although the procedures used are referred to generally as decolourizing, the term implies a complete removal of colour, whereas in practice the aim is very often to produce a very slight but reproducible tint showing a shade of pink or blue according to the particular preference of the company whose goods are to be sold in the type of container in question.

The need for decolourizing arises from the presence of col-
our-producing impurities, in particular iron oxide, in the raw
materials. Most of the iron oxide is introduced via the sand,
a typical sand used in container manufacture containing 0.03%
Fe_2O_3. It is not economic to avoid the problem by using purer
raw materials.

If no steps were taken to control the colour, the glass
melted in a typical oil- or gas-fired tank furnace would have a
pronounced blue tint. About 80% of the iron would be present
as Fe^{3+} and the rest as Fe^{2+}. Both ions have their strongest
absorptions outside the visible, the Fe^{2+} ions in the infra-red
and the Fe^{3+} in the ultra-violet. However the absorption bands
are broad and both ions contribute a colour to the glass, that
due to Fe^{2+} being blue and that due to Fe^{3+} being yellowish green.
In the central region of the visible spectrum where the eye is
most sensitive, the molar absorptivity of the Fe^{2+} is approximately ten times greater than that of the Fe^{3+} ion. This provides the basis for one of the methods of decolourizing, which
is to add materials to the batch which oxidize the ferrous iron
to ferric. This is often referred to as chemical decolourizing.
Clearly any material added to produce this effect must not itself
colour the glass.

The other principal method of decolourizing, which may be
used alone or in combination with the chemical method, is to add
materials which produce a colour which neutralizes that due to
the iron, i.e. its absorptivity, when added to that due to the
iron present, gives a net absorptivity which is practically constant throughout the visible spectrum.

The detailed practice used in the industry tends to vary
for a number of reasons. New raw materials become available
from time to time which are found to accelerate the rate of melting. Their introduction may require some change in decolourizing practice. A relatively recent example of this was the introduction of the use of a selected grade of blast furnace slag,
sold under the trade name "Calumite". Another factor is that
the decolourizing additives fluctuate in price and this can influence the selection of the materials to be used. Also a glass
manufacturer who has to use a grade of sand of varying iron content obviously has to vary the quantity of decolourizing materi-

als used to maintain a reproducible tint.

A good impression of the wide range of materials used for decolourizing is given in a paper by Herring et al. (1970). It is enough to say here that probably the most commonly used materials for chemical decolourizing are arsenious oxide, As_2O_3, which is usually used in combination with sodium nitrate, or a material containing a high percentage of CeO_2. Both these materials are effective in oxidizing Fe^{2+} ions in glass to Fe^{3+} and neither produces any noticeable colour in any of their common valence states (As^{3+} or As^{5+} in one case, and Ce^{3+} or Ce^{4+} in the other). The most commonly used materials for physical decolourizing are selenium or CoO, the former producing a pink colour, and the latter a blue.

Decolourizing with the use of selenium is not easy to control, especially if chemical decolourizers are used also. The material is volatile and is also easily oxidized by chemical decolourizers if these are present. Furthermore the development of the selenium pink is affected by the heat treatment which the glass receives as it passes through the manufacturing process.

H. Laser Glasses

The year following the discovery by Maiman of the ruby laser, Snitzer (1961) reported the laser action of a potassium barium silicate glass containing 2 weight per cent of neodymium oxide, Nd_2O_3. A tremendous amount of work followed to investigate the laser action of a number of the rare earth ions (especially Er^{3+}, Yb^{3+} and Nd^{3+}) in a wide range of glass compositions and single crystal materials. Review articles by Snitzer (1966, 1973) and Young (1969) and the book by Patek (1970) deal specifically with the work done on glass lasers.

The characteristics of a glass laser material containing a particular active ion, say Nd^{3+}, differ from those of a single crystal containing the same ion, the differences depending very largely on the fact that the former host material has a random structure whilst that of the latter is periodic. Consequently the energy levels responsible for the laser action in a Nd^{3+} glass laser are spread over a much wider range than in a single crystal material. This has some practical advantages and some

230

disadvantages, as a consequence of which glass laser materials
are preferred for some applications whilst single crystal materials are superior for others.

Before considering this aspect of the different characteristics of glass and crystalline laser materials, it is worth
mentioning two simple points which also have an important bearing on the choice of material. An important advantage of glass
for lasers is that it is relatively easy to produce a wide range
of shapes in material of high optical quality. Thus laser glass
has been made, and used, in forms ranging from fibres to discs
measuring tens of centimetres in diameter. The largest pieces
of laser glass in use are probably the amplifier discs in the
laser fusion facility at the Lawrence Livermore Laboratory in
the U.S.A. It would be enormously expensive and probably tech-
nically impossible to make single crystals of a similar size.
In any case, this happens to be an application for which glass
is the preferred material on account of its laser characteristics
A basic disadvantage of glass is that oxide glasses have lower
thermal conductivities than oxide crystals. In high power applications where a good deal of heat has to be dissipated this
presents quite a serious problem and limits the laser power.

In the normal configuration of an optically pumped solid
state laser, the laser material in the form of a cylindrical
rod is aligned between two plane parallel mirrors of reflectivity R_1 and R_2 (Fig. 127). R_1 is usually 100% whereas R_2 is

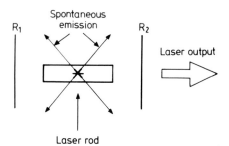

Fig. 127. Schematic diagram of a solid laser (Snitzer, 1973).

only a few percent and it is through mirror 2 that the laser
beam emerges. The axis of the rod is normal to the mirror surfaces. The rod is strongly irradiated by pulses of light from
one or more high intensity flash lamps disposed around the rod,

mirrors being used to concentrate the light onto it.   Only a
few percent of this light leaves the system in the laser beam.
The first requirement of a good laser material is that it should
be able to absorb as much as possible of the incident light in
a useful way, exciting the Nd^{3+} ions (say) to higher energy levels.   The energy levels of the Nd^{3+} ion are shown in Fig. 128.

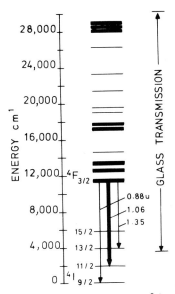

Fig. 128.   Energy level diagram for Nd^{3+} in a barium
crown glass.   (Snitzer, 1964).

A characteristic of the absorption spectra of the rare earth ions
dissolved in a transparent material is a large number of absorption bands spread over the wavelength range from the near UV to
the near IR.   These bands are much narrower than those due to
the d → d transitions of the transition metal ions considered
earlier in this chapter.   The narrowness arises from the fact
that the rare earth ions have an incomplete inner shell of electrons which is shielded from the effects of ligand oxygen ions
to a greater extent than the d electrons in the outer shell of
the transition elements.

Light energy is thus pumped into the laser material.   The
excited ions lose some energy without the emission of radiation,
their energies falling to the level labelled $^4F_{3/2}$.   The transition which is responsible for the most used laser action is from
this level to that labelled $^4I_{11/2}$.   This transition *is* accom-

panied by the emission of radiation. The emitted radiation
lies within a relatively narrow band in the near infra-red cen-
tered around a wavelength of 1.06 micron (Fig. 129).

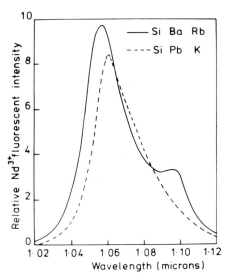

Fig. 129. The relative fluorescence of Nd^{3+} in various
laser glasses at 300°K (Snitzer, 1973).

Following an excitation pulse, it takes a time of the order
of milliseconds for the substantial completion of the de-excita-
tion from the upper to the lower laser level. Thus the emission of light takes place within a very short period of time.
The time is markedly affected by the composition of the host
material and it is an important parameter affecting laser performance.

The emission of radiation from a material of wavelength
longer than that of that of the exciting radiation is known as
fluorescence. The phenomenon had been studied in both crystalline and glassy materials for many years before it came to be
exploited in lasers. In fluorescence the light is emitted in
all directions. It requires the very special geometrical arrangement of a laser to funnel this emission into a very narrow
range of angles, which is one of the best known features of
laser light. Other well known features of laser light are its
high intensity and its monochromaticity.

The main difference between the fluorescent emission from
a glass containing a rare earth ion and that from a crystal is

that in the glass the emission band is much broader (by a factor
of 20-30) than in the crystal.   This is a consequence of the
greater variability of the structural environment of the lasing
ion when dissolved in a glass.   The consequences of this broader emission line are considerable.

It would be inappropriate here to give even the simplest
explanation of laser action.   Some topics are so complicated
that the result of attempting to explain them in a paragraph is
usually unsatisfactory, if not misleading.   Lasers probably fall
in this category, particularly when one is attempting to explain
respects in which one type of laser differs from another.

However, it is perhaps worth going a little further in an
attempt to explain the consequences of the broader emission line
in the glass laser.

Laser action depends upon the amplification of light which
is being propagated within a very narrow range of angles close
to the rod axis.   Conditions for amplification are satisfied
only within this narrow range of angles and only within a very
narrow range of wavelengths close to that at which the emission
band has its greatest intensity.   The physical phenomenon on
which lasing depends is that a photon having a frequency corresponding to that of the difference in energy between level $^4F_{3/2}$
and $^4I_{11/2}$ can stimulate or trigger off the transition between
these two levels.   This too has been known for many years and
much of the relevant basic physics of stimulated emission was
analyzed by Einstein in the early years of the century.   Thus
one photon can be midwife to another.   The delivery is rapid
and the population growth can be very large.   However there are
competing processes.   A massacre of the innocents is taking
place.   Many of the photons propagating through the laser rod
may be absorbed if the material contains impurities.

In glasses, ferrous iron is particularly objectionable as
an impurity since it has a peak absorption close to the laser
wavelength.   Equally important, some photons are lost in stim-
ulating the reverse transition from $^4I_{11/2}$ to $^4F_{3/2}$ and indeed
laser action is only possible when the previous excitation of
the laser material by the flash lamps has resulted in a greater
number of Nd^{3+} ions in the $^4F_{3/2}$ state than the $^4I_{11/2}$ state

(a so-called "inverted population.")

To achieve an inverted population there is a minimum amount
of energy which has to be fed in during the exciting flash.
This minimum is referred to as the threshold.   Unless this is
exceeded laser action cannot occur.   It can be shown that the
threshold energy is inversely proportional to the breadth of
the fluorescent band.   Consequently glasses are worse than
crystalline hosts from this point of view.   Higher threshold
energies are required to operate a glass laser.

Yet the same feature, i.e. the breadth of the emission band,
is responsible for one of the most important advantages of glasses
as a laser material.   Remember that a photon can only stimulate
emission at its own wavelength.   When the emission band is narrow, as in a crystal, a photon stimulates emission from a much
higher proportion of the lasing ions than in a glass, where the
transitions are spread over a much wider range of energies.
The midwife photon in the crystal is much more effective than
in the glass where a greater variety of mothers have to be dealt
with.   The higher birth rate in the crystal rapidly decreases
the number of pregnant mothers and the midwives soon find themselves unemployed and laser action ceases.   It is better in
applications where large pulses of energy are to be delivered
to have available a substantial reservoir of slightly more difficult births which can make themselves available for the production of offspring in a rather less enthusiastic way.   This
is where glasses have the advantage.   The upper laser levels
are not so rapidly depopulated.   Energy can be more easily
stored in the upper laser levels and high energy pulses delivered.

Thus glasses have the advantage in applications requiring
the storage of energy and the delivery of pulses of laser light.
These applications are, for example, in the fields of welding
of components, in laser range finders and, most spectacularly,
in research on the production of power by laser fusion where
huge quantities of energy have to be delivered in very short
pulses on hydrogen isotopes.   To illustrate the versatility of
glass, it is worth noting that these energy pulses, delivered
with the aid of large discs of laser *glass*, are concentrated
on the hydrogen isotopes which are contained with minute *glass*
microspheres.

The reader who has found some of the gynaecological analogies in this section rather laboured, would be advised to consult the clear, clinically mathematical but nevertheless relatively simple accounts of laser physics given in the books by Lengyel (1966) and Ditchburn (1976) before attempting to follow the accounts of work on laser glasses in the references given at the beginning of this section.

CHAPTER VIII

ELECTRICAL PROPERTIES

A. Introduction

The construction of the first successful filament lamps by
Edison and Swan in the late nineteenth century and the later development of vacuum diode and triode valves led to the rapid
growth of two new industries, the electric lamp industry and the
electronics industry. As their products increased in complexity,
needs arose for new glasses for lamp and valve envelopes and also
for internal components. Consequently there exist today a number of large glass companies and glass manufacturing divisions
of companies in the electrical industry, which supply the glasses
and glass components required.

Although the need for a large range of glass types is perhaps not so great as in the optical industry, the requirements
of the electrical industry have had a great influence in stimu-
lating the search for new glass-forming systems and the study of
relationships between glass properties and glass composition.

Practically all the glasses used commercially are good electrical insulators at room temperature, and it is only in a relatively small number of applications that the electrical properties are of critical importance. Thus the glass used in the
"pinch seal" of a filament lamp must have a particularly high
resistivity at elevated temperatures, since the wire leads which
are sealed through the glass are only a few millimetres apart,
the voltage between the wires is relatively high and the pinch
is at a temperature of about 200°C when the lamp is running.
Another application where exceptionally good electrical properties are required (in this case a low dielectric loss) is in the
output window of microwave valves which may have to transmit several kilowatts of power at centimetre wavelengths.

On the other hand there are a number of situations in which
practical use is made of the fact that the resistivity of most
glasses falls to quite low values at high temperatures. A sufficiently high current can then be passed through the glass to
make it possible to melt the material electrically, the current
passing between molybdenum or semi-conducting tin oxide electrodes which are submerged in the glass. Although electric

melting is economically attractive in only a few countries, it
is becoming more common.   Also electric boosting is widely used
to supplement the main heating provided by oil or gas flames
(Stanek, 1977).   Electric heating can also be used to seal together glass components, a practice which is widely used in the
manufacture of black-and-white TV tubes.   The current is led
into the glass through gas burners, the hot gases of the flame
acting as current-carrying "brushes".   The flames also serve
the purpose of heating the glass to a sufficiently high temperature for it to become conducting (Guyer, 1969).

Thus there are many technological reasons for interest in
the electrical properties of glass.   This interest has been
greatly stimulated over the last twenty years by the discovery
of semi-conducting glasses and of high-speed switching effects
that can be produced in some of these materials (Pearson et al.
1962; Ovshinsky, 1968; McMillan, 1976).   These discoveries
directed the attention of solid-state physicists to the problem
of understanding electronic conductivity in amorphous solids, a
topic of considerable scientific interest.

Because of these recent developments, the literature on the
electrical properties of glasses is expanding rapidly.   Fortunately a number of good review articles exist.   Owen (1963, 1977)
and Hughes and Isard (1972) deal with the electrical properties
of ionically conducting glasses whilst the reviews of Owen (1970,
1977), Owen and Spear (1976) and Mott (1968, 1977) are concerned
with the semiconducting materials.   There is also a useful bibliography of this field, published by the International Commission on Glass (1977).

B.   Ionically conducting glasses

1.   Main Features of the Conduction Process

The conduction of electricity in most oxide glasses is due
to the positively charged alkali ions moving under the influence
of the applied field.   The energy which an ion requires to overcome forces of attraction which tend to hold it at one position
in the structure is supplied from the thermal energy in the material.   Consequently as the temperature of the glass is in-
increased, a greater proportion of the ions is enabled to move and
the conductivity increases.   This effect will be dealt with in

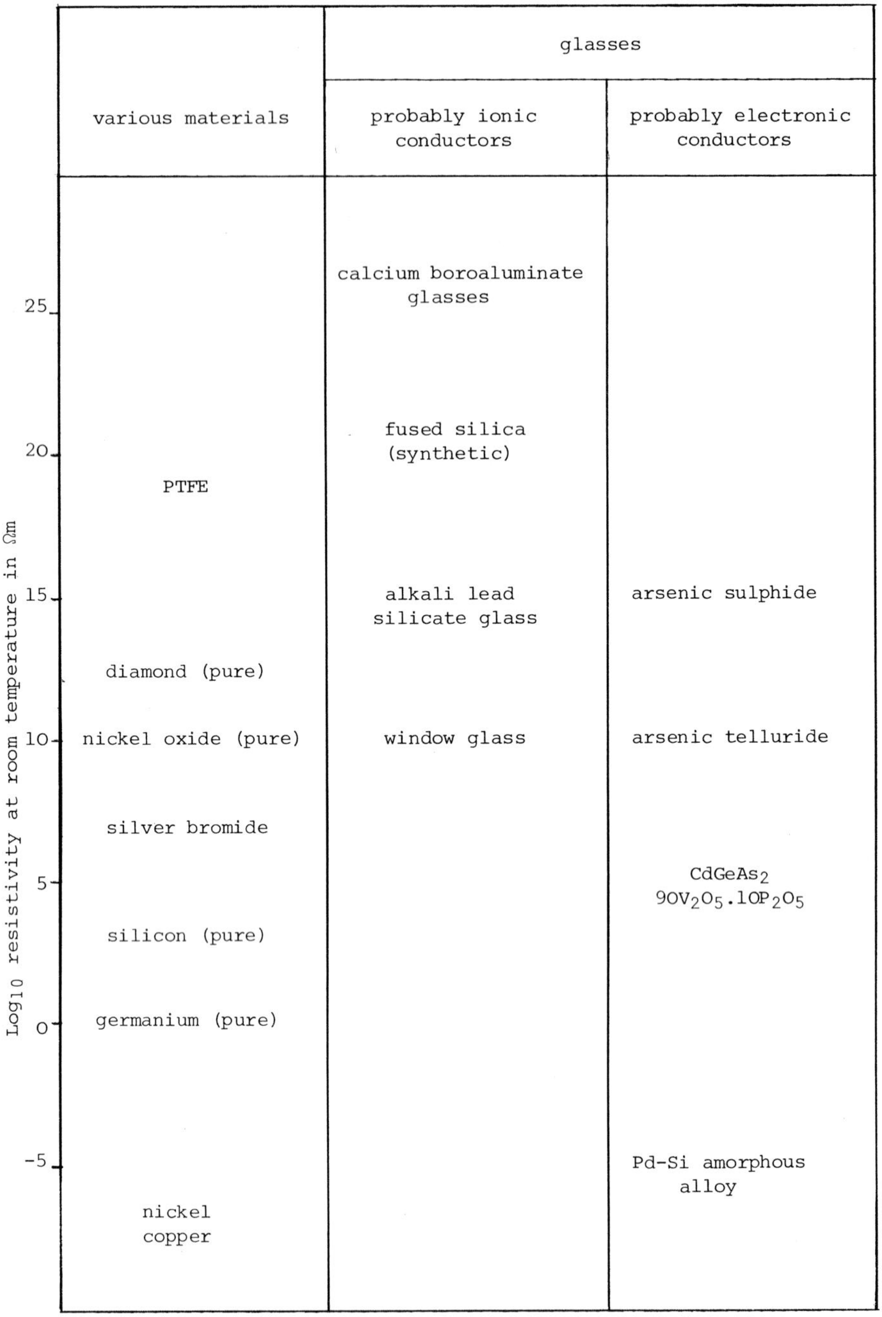

| various materials | glasses | |
	probably ionic conductors	probably electronic conductors
	calcium boroaluminate glasses	
PTFE	fused silica (synthetic)	
diamond (pure)	alkali lead silicate glass	arsenic sulphide
nickel oxide (pure)	window glass	arsenic telluride
silver bromide		
silicon (pure)		CdGeAs$_2$ 90V$_2$O$_5$.10P$_2$O$_5$
germanium (pure)		
nickel copper		Pd-Si amorphous alloy

Log$_{10}$ resistivity at room temperature in Ωm

25 — 20 — 15 — 10 — 5 — 0 — -5 —

Fig. 130.   Resistivities at room temperature.

more detail in the next section.

Figure 130 shows that glasses can be made with specific resistivities at room temperature ranging from less than 10^{-5} ohm
m to more than 10^{25} ohm m. Although all the commercial oxide
glasses are good insulators, the metallic alloy glasses conduct
electronically and have conductivities of the same order as crystalline metals. In between these two extremes are the oxide
and chalcogenide semi-conducting glasses.

Figure 131 shows how rapidly the resistivities of commer-

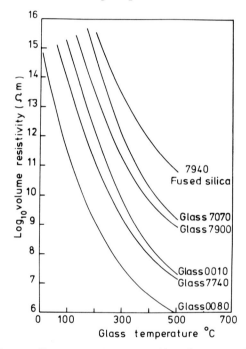

Fig. 131. Effect of temperature on the d.c. resistivity of
some commercial glasses: 7070 - borosilicate; 7900 - 96%
silica; 0010 - potash lead silicate; 7740 - borosilicate;
0080 - soda-lime silicate (Corning Glass Works).

cial glasses decrease with temperature. As the alkali cations
move towards the negative electrode, the layer of glass in contact with the anode suffers a reduction of alkali content. The
resistance of this layer rises so that the resistivity of the
material appears to increase. This polarization effect is
troublesome when using d.c. methods for measuring resistivity,
unless non-polarizing electrodes are used which are capable of

supplying alkali ions at the anode-glass interface. Understand-
ably, when melting glass electrically it is necessary to use alternating current.

An interesting experimental demonstration of the fact that
electrical conduction in a soda-lime-silica glass is due almost
entirely to the motion of the sodium ions has been described by
Burt (1925). In this experiment metallic sodium is produced by
electrolysis through the wall of an incandescent filament lamp
bulb. (Anyone wishing to carry out this demonstration should
note that the lamp should be of the vacuum type, not the normal
commercial type which has a gas filling.) Electrons liberated
from the hot filament neutralize the sodium ions at the inner
surface of the bulb. A continuous supply of sodium ions at the
outer surface of the bulb is provided by partly immersing the
bulb in a fused salt bath of low liquidus temperature containing
sodium ions. Applying a d.c. voltage between the salt and the
filament drives the sodium ions through the wall of the bulb.
By relating the increase in weight of the bulb, i.e. the weight
of sodium which has been transported through the glass, to the
quantity of electricity which has been passed through the circuit, Burt showed that within experimental error, all the current had been carried by the sodium ions.

For investigating the nature of the current carriers in
glasses of various compositions, the Tubandt disc method (Fig.
132) has frequently been used. The three glass discs, all of

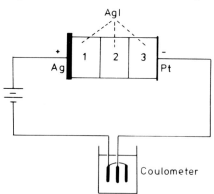

Fig. 132. Tubandt disc experiment for measuring transport
numbers in α-AgI (Hughes and Isard, 1972).

the same composition, are weighed before and after a measured
quantity of electricity has been passed.   Only the discs in
contact with the electrodes should show a weight change.   From
the measurements it is possible to calculate the fraction of the
current carried by alkali ions and that carried in some other
way, possibly by electrons or protons neither of which contribute significantly to the weight change.   Figure 133 shows re-

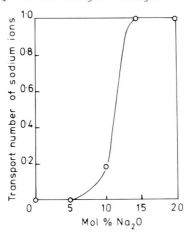

Fig. 133.   Sodium ion transport numbers for glasses in the
system xNa$_2$O,$(1-x)$PbO.SiO$_2$ (Hughes and Isard, 1972).

sults for the variation with Na$_2$O content in a series of Na$_2$O-
PbO-SiO$_2$ glasses of the transport number of the sodium ion, i.e.
the fraction of the current carried by that ion.   In the high
alkali compositions practically all the current is due to sodium
ion movement but below 5% Na$_2$O other current carriers are involved.

2.   A Simple Model of the Ionic Conduction Process

The theory described in this section is also used to account for the particular way in which the conductivity of crystalline materials such as sodium chloride varies with the temperature.   The fact that it is also applied to glasses is an
indication of the relative insensitivity of the process of ionic
conduction to the degree of regularity of the structure through
which the ions move.

Figure 134a is a schematic representation of the variation
with position of the energy of a cation as it moves along one

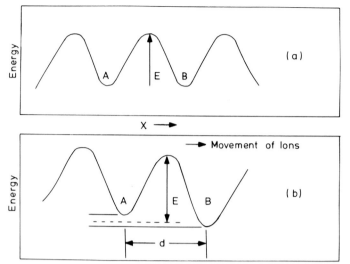

Fig. 134. Schematic diagram of energy barriers (a) before
and (b) after applying an electric field (Owen, 1963).

direction in a crystal. The potential energy is lowest when
the cation is symmetrically surrounded by anions and it passes
through a maximum as the cation moves from one such position to
the next. In making this movement work must be done to overcome the forces of attraction between the cation and surrounding
anions and the forces of repulsion between the cation and other
cations to which it comes closer in its movement. For a cation
to be able to move, an adjacent vacant cation site must be available for it to move into. This represents the situation when
no electric field is applied to the material. The cation has
identical potential energies in the sites A and B.

Even in the absence of an electric field, ions are moving
from one low energy position to the next. At a temperature $T^{\circ}K$
the probability that a vibrating ion has an energy greater than
E, and is thus able to move from A to B or vice versa, is proportional to $\exp(-E/kT)$ where k is the Boltzmann constant. If
the ion is vibrating about its equilibrium position at a frequency ν, which is typically in the range 10^{13} to 10^{14} Hz, each
ion will make a number of jumps per second equal to $\nu \exp(-E/kT)$.
In a given period of time it will make the same number of jumps
in each direction and consequently there will be no net movement
of charge, so the current through the material is zero.

Application of an electric field has the effect of tilting the energy diagram (Fig. 134b). Since the net work done in moving a univalent ion against the applied field ξ is ξed where e is the electronic charge and d is the distance between adjacent sites, it follows that the bottom of one energy well must be that much higher than the one immediately to its right. An equivalent way of describing the situation is to say that the energy barrier on the left of each well has been increased by $\xi ed/2$ and that onthe right has been decreased by the same amount. Consequently the frequency of jumps to the right will be increased and that to the left decreased. The current density, I, is proportional to J, the net number of jumps per second made by each ion in the direction of the field. J is given by

$$J = \nu \exp(-(E-\xi ed/2)/kT) - \nu \exp(-(E+\xi ed/2)/kT)$$
$$= \nu \exp(-E/kT)(\exp(\xi ed/2kT) - \exp(-\xi ed/2kT)) \quad (99)$$

The relation between I and J is $I = nedJ$ where n is the number of cations per unit volume. For field strengths normally encountered, the value of ξed is very small compared to kT. Consequently the equation for I may be simplified to give

$$I = (ne^2 \nu \xi d^2/2kT).\exp(-E/kT). \quad (100)$$

The electrical resistivity is readily shown to be

$$\rho = (2kT/ne^2 \nu d^2).\exp E/kT \quad (101)$$

Taking the logarithms of both sides of this equation

$$\ln\rho = \ln(2kT/ne^2 \nu d^2) + E/kT \quad (102)$$

The variation with temperature of the first term on the right hand side is very small compared with that of the second. Thus the equation has a similar form to the so-called Rasch-Hinrichsen equation, $\log\rho = A + B/T$, which describes reasonably well the variation of resistivity with temperature for many ionic crystals and oxide glasses (Fig. 135). In the equation A and B are constant for a particular material.

The value of E, the activation energy for the conduction process in the glass is readily obtained from the slopes of the lines in the figure. The values obtained for those commercial oxide glasses which contain sodium ions are usually in the range 18-20 k cal mole^{-1}. (Note that the mathematical model given above has been developed in terms of the energy per ion. Activation energies for rate processes such as ionic conduction are

244

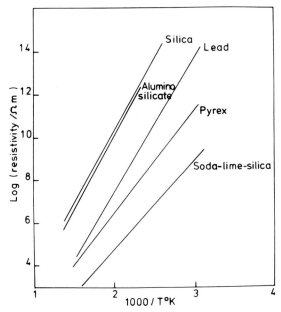

Fig. 135. Variation of log resistivity with 1/T°K for a
number of commercial glasses. (Holloway, 1973).

usually quoted as the energy per mole. This value is N times
greater than the value of E in Equation 102. N is Avogadro's
number.)

Because the structure of a glass is a random one, the potential energy diagram must be more complicated than that in Fig.
134, possibly somewhat like that in Fig. 136. For this there

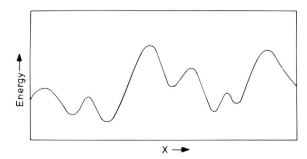

Fig. 136. Schematic diagram showing the distribution of
energy barriers which may exist in a glass. (Owen, 1963).

is a range of energy values for the energy barrier. The value
of E determined from the measurements is therefore likely to be
some kind of average value.

3. Effects of Glass Composition

(a) *Silica and binary silicate glasses*.   Commercial grades of
vitreous silica contain different amounts of impurities which
markedly affect the electrical properties.   At one time nearly
all this material was made by the vacuum fusion of natural quartz
and some grades still are.   They contain various amounts of metallic oxide impurities, including alkali oxides.   The more recently developed synthetic material made by pyrolysis of silicon
tetrachloride is very much purer.   Resistivities of the various
grades measured at 300°C differ between grades by factors as
large as 10^4 (Owen and Douglas, 1959).

In the binary alkali silicate systems, the resistivity decreases monotonically with increasing R_2O content (Fig. 137).

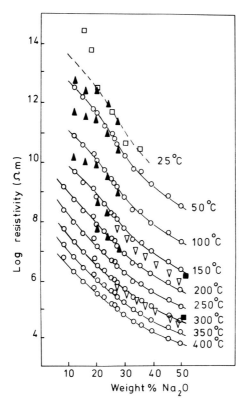

Fig. 137. Variation of resistivity with composition for
glasses in the system Na_2O-SiO_2. (Hughes and Isard, 1972).

Although the results of various workers agree very well at high

246

alkali contents, there are considerable discrepancies at low alkali contents.   This may simply be a consequence of the difficulty of making sufficiently homogeneous samples.

The increase in conductivity with increasing R_2O content is
not simply proportional to the increasing number of alkali ions
per unit volume of the glass.   Thus an increase in alkali content from 15 to 45 mol% increases the conductivity approximately
a hundredfold.   The significant change causing the increase in
conductivity is the marked reduction in activation energy (Fig.
138).

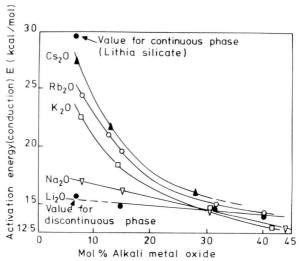

Fig. 138.   Variation of activation energy for conduction
with composition in the alkali silicate glass systems.
(Charles, 1966).

The comparison in Table XXIII of the resistivities of the various alkali silicate glasses at the same molecular composition
and the same temperature shows that the values increase with increasing cation size.   However there is no simple relationship
between activation energy and cation size which holds over the
whole range of composition.   Figure 138 shows that the activation energies vary in a complicated way with composition, the
curves for the various alkali systems crossing between 30 and
40 mol% R_2O.   The variation of activation energy with temperature (Fig. 139) is also markedly different for the various alkali silicates.

TABLE XXIII   (Charles, 1966)

d.c. Resistivity at 100°C (ohm-cm)

Mol% Alkali	Li_2O	Na_2O	K_2O	Rb_2O	Cs_2O
15	1.7×10^7	1.4×10^8	3.1×10^9	1.4×10^{10}	3.3×10^{11}
30	4.2×10^6	3.8×10^6	6.5×10^6	9.7×10^6	1.6×10^8
40	9.7×10^5	1.4×10^5	9.4×10^5	3.0×10^6	–

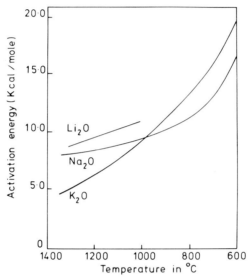

Fig. 139. Variation of the activation energy for conduction
with temperature for alkali silicate melts of composition
30 mol% $R_2O.70\%$ SiO_2. (Tickle, 1967).

Authors who have studied the resistivity of glasses for scientific rather than technological reasons have been more interested in the values of the activation energies than the resistivity values themselves, because the observed changes in activation energy with glass composition and temperature may eventually
lead to a better understanding of the conduction process or to
an insight into changes of glass structure with composition.
Attempts to draw conclusions about the nature of the conduction
process from the values of specific resistivity are not in any
case likely to be successful. Increasing the alkali content
not only increases the proportion of current-carrying ions in

the composition, it also changes the density of the glass. To
eliminate effects of density and alkali concentration it is necessary to calculate the value of the equivalent conductance Ω,
i.e. the conductivity of the glass.converted to a concentration
of one gram ion of alkali per litre. Ω is calculated from $\Omega =$
$1/\rho c$ (where ρ is the specific resistivity and c is the alkali
concentration in gram ions per litre) and can be related in a
simple way to the cation velocity, v, by the equation $\Omega = Fv$
where F is Faraday's constant. Few authors have expressed
their results in this way, a notable exception being Tickle (1967).
Figure 140 shows some of his results for the equivalent conduc-

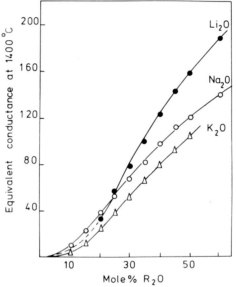

Fig. 140. Variation with composition of the equivalent
conductance at 1400°C of alkali silicate melts. (Tickle, 1967).

tance of alkali silicate melts at 1400°C. One would need to
have results at lower temperatures plotted in the same way to
judge whether there appeared to be any significant pattern in
the results.

An additional factor complicating the interpretation of
measurements made below the liquidus temperature is the possibility of phase separation. Charles (1966) has shown that the
electrical properties of Li_2O-SiO_2 and Na_2O-SiO_2 glasses are
markedly affected by metastable immiscibility, the glasses sep-

arating, to an extent depending on composition and heat treatment, into a high conductivity and a low conductivity phase. This may explain the discrepancies in the measurements at low Na_2O contents which were referred to earlier. No such separation effects occur in the K_2O-SiO_2 and Rb_2O-SiO_2 systems.

Although immiscibility does not occur in the alkali silicate melts above the liquidus temperature it may be significant that the more marked changes in slope of the equivalent conductance-composition curve in the Li_2O-SiO_2 system is for that system which shows the largest deviation from ideality. One would expect a greater tendency for clustering of the cations in the Li_2O-SiO_2 melts, these clusters being surrounded by regions of lower than average alkali content. This may be the reason for the comparatively low value of the equivalent conductance of the Li_2O-SiO_2 melts between O and 30 mol% Li_2O.

(b) *The mixed alkali effect* This is perhaps the most remarkable of all the effects of glass composition on properties. Figure 141 shows the electrical resistivity results of Charles

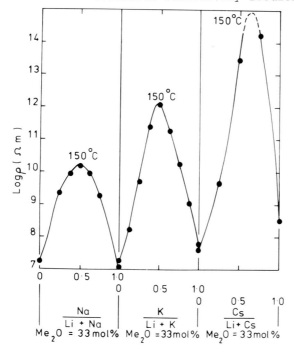

Fig. 141. Resistivity data for mixed alkali silicate glasses containing 30 mol% R_2O. (Charles, 1965).

(1965) for three mixed alkali systems, in all of which the molecular percentage of total alkali is constant.   In each system
the resistivity passes through a pronounced maximum as one alkali ion is substituted for the other.   The activation energy
also passes through a maximum at the same composition.   Charles
suggests that in these particular systems the effect may be due
to metastable immiscibility.   However he recognizes that this
explanation cannot apply to all the systems in which the effect
is observed, e.g. the $Rb_2O-Cs_2O-SiO_2$ system which shows no metastable immiscibility.   The effect is also shown in results obtained above the liquidus (Tickle, 1967).

The many theories which have been proposed to explain the
effect have been reviewed by Isard (1969).   The most recent
theory by Hendrickson and Bray (1972) proposes that it arises
from an energy of interaction between dissimilar cations on adjacent sites.   The interaction arises because of the different
natural frequencies of vibration of the cations.   The theory
predicts a mixed alkali effect in glasses containing different
isotopes of the same alkali.   The observation of such an effect
by Abou el-leil et al. (1978) encourages one to believe that the
effect has at last been given a satisfactory explanation.

The effect is exploited in glasses of high electrical resistivity.   Thus the 30 wt% PbO glass used in making the pinch
seal in incandescent filament lamps is a mixed alkali glass containing both K_2O and Na_2O.

(c) *Effects of a third oxide*. Apart from the studies of the
mixed alkali glasses, relatively little systematic work has been
done on the electrical resistivity of three component silicate
glasses.   Much of the information available is to be found in
the book by Mazurin et al. (1975).

Figure 142 shows results of Mazurin and Brailovskaya (1960)
for a series of glasses having the general composition 20 mol%
Na_2O, 20% RO, 60% SiO_2, all the measurements having been made
at 150°C.   The results were interpreted simply in terms of the
divalent ions blocking the motion of the alkali ions through
the glass network.   These ions occupy similar sites in the
glass structure to the alkali ions and it is suggested that the
larger the divalent ion, the greater will be the blocking effect and hence the higher the resistivity.

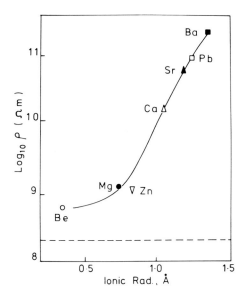

Fig. 142. The effect of the divalent ion radius on the resistivity of Na_2O-RO-SiO_2 glasses containing 20 mol% R_2O and 20 mol% RO. The dashed line represents the resistivity of a glass of composition 20 mol% Na_2O, 80% SiO_2. (Owen, 1963).

At high temperatures this size effect disappears. The results of Katanyan et al. (1965) in Figure 143 show that the re-

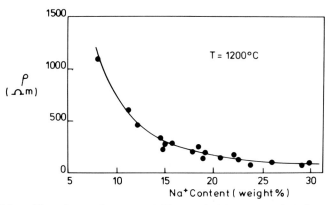

Fig. 143. The dependence on the weight percent of sodium ions in the glass of the resistivities of three component silicate glasses. The data points represent glasses containing MgO, CaO, BaO, BeO and CdO with between 68 and 78 weight% SiO_2. (Hench and Shaake, 1972).

sistivity of many three-component glasses containing CaO, MgO,

BaO, BeO and CdO depends only on the alkali content of the glass.

The effect of varying the Al_2O_3 content of glasses in the system Na_2O-Al_2O_3-SiO_2 shows some interesting features. Figure 144 shows the variation of activation energy with alumina con-

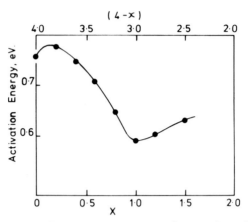

Fig. 144. Effect of composition on the activation energy for d.c. conduction of glasses in the system $Na_2O.xAl_2O_3.2(4-x)SiO_2$. (Isard, 1959).

tent for a series of glass compositions represented by the general formula $Na_2O.xAl_2O_3.2(4-x)SiO_2$ (Isard, 1959). These results should be considered in conjunction with the refractive index results given in Chapter VI for glasses in the same ternary system. In the region of compositions where x is less than 1, all the Al^{3+} ions enter the silicate network, each such ion being four-fold coordinated by oxygen anions. This results in a decrease in the number of non-bridging oxygens as the Al_2O_3 content increases. Isard suggests that this will have the effect of expanding the coordination shell of the oxygens around the sodium ions, which in turn will give rise to the observed decrease in activation energy. This type of structural change is expected to cease at x = 1, i.e. at the composition $Na_2O.Al_2O_3.$ $6SiO_2$. Further additions of Al_2O_3 result in the Al^{3+} ions entering modifier positions in the glass network when they are six-fold coordinated by oxygen. This is accompanied by an increase in the number of non-bridging oxygens, a decrease in size of the coordination shell around the sodium ions and a consequent rise in the activation energy. Somewhat similar changes

have been observed in the system Na_2O-GeO_2 and these have also
been attributed to changes in coordination number (Ivanov, 1963;
Ivanov et al., 1965).

Work on the electrical properties of other oxide glasses,
e.g. the borate and phosphate glasses, is even scantier than on
the silicate glasses. The reader is referred to the review articles quoted in the introduction to this chapter.

C. Semi-conducting Glasses

1. Types of semi-conducting glasses and their charac-
teristic features

Although it is less than 20 years since it became evident
that a wide range of glasses could be made which conduct electronically rather than ionically, the literature on the semi-
conducting glasses is now considerably larger than that on the
glasses dealt with in the previous section.

The reasons for this intensive activity are partly scientific and partly technological. There already existed for crystalline solids a well developed theory for the band structure
of the electron energy levels and for the mechanisms of electrical conductivity. It has been a challenge to extend this
theory to materials having structures without long range order
(Mott, 1978). Considerable progress has been made in this direction.

The technological interest was for some years centred on
the study of the high speed switching phenomena observed in simple devices which can be made from some of the semi-conducting
glasses. At the time of writing it appears that this interest
is waning, partly because it has been found difficult to make
the devices with a sufficiently reproducible behaviour, and
partly because the development of other high speed switching
devices has made the glass switches less attractive. However
there is now a growing interest in the possibility that amorphous, if not glassy, semiconductors may find an application in
making relatively cheap solar energy converters. One semi-conducting glass, vitreous selenium, has already been used for
many years in xerographic copying machines.

There are two groups of semi-conducting glasses, one based
on the chalcogen elements (S, Se and Te) and the other on com-

positions containing high percentages of one or more of the transition metal oxides (especially V_2O_5, MoO_3 and WO_3). The ranges of glass-forming compositions within each group are wide.

Most of the binary and ternary chalcogenide glasses are made by melting together the chalcogen element with elements from groups IV and V of the periodic table, in particular Si, Ge, P, As and Sb. Figure 145 gives a general picture of the

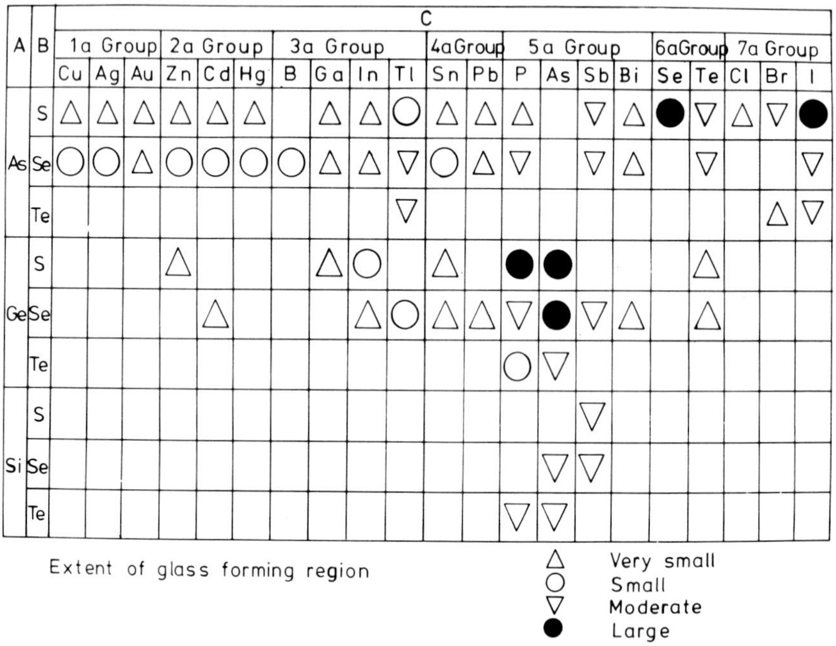

C — groups: **1a Group** (Cu, Ag, Au); **2a Group** (Zn, Cd, Hg); **3a Group** (B, Ga, In, Tl); **4a Group** (Sn, Pb); **5a Group** (P, As, Sb, Bi); **6a Group** (Se, Te, Cl); **7a Group** (Br, I)

A	B	Cu	Ag	Au	Zn	Cd	Hg	B	Ga	In	Tl	Sn	Pb	P	As	Sb	Bi	Se	Te	Cl	Br	I
As	S	△	△	△	△	△	△		△	△	○	△	△	△		▽	△	●	▽	△	▽	●
As	Se	○	○	△	○	○	○	○	△	△	▽	○	△	▽		▽	△		▽			▽
As	Te										▽									△	▽	
Ge	S				△				△	○		△		●	●			△				
Ge	Se					△			△	○		△	△	▽	●	▽	△	△				
Ge	Te											○	▽									
Si	S															▽						
Si	Se														▽	▽						
Si	Te													▽	▽							

Extent of glass forming region
△ Very small
○ Small
▽ Moderate
● Large

Fig. 145. Glass formation in ternary chalcogenide systems: A-B-C. (Owen, 1970).

ease of the glass formation in various ternary systems. Since the constituents are volatile or react readily with oxygen, the glasses cannot be made by melting in an open crucible. The mixture of elements in powder form is placed in a thick-walled vitreous silica tube which is then evacuated and sealed off. The tube is heated in an electric furnace, usually equipped with a rocking mechanism so that the melt can be agitated to promote homogeneity. Although the melts are relatively fluid, those containing silicon as a constituent require heating for several hours at temperatures of about 1000°C before all the silicon is incorporated in the melt. At the end of the melting period the tube may need to be rapidly cooled to prevent the glass from crystallizing.

Before 1960 most of the work on the chalcogenide glasses had been carried out in the Soviet Union, especially by Kolomiets and his colleagues. A summary of this early work has been given by Owen (1963). During the last ten years interest has become world-wide and several excellent review articles are available, e.g. Owen (1970), Weiser (1976).

The first publications describing the semi-conducting properties of glasses based on the transition metal oxides were those of Denton et al. (1954) and Baynton et al. (1956), Both describe work on glasses containing high percentages of V_2O_5 and much of the more recent work has been concerned with these **vanadate glasses.** They can usually be melted at temperatures below 1000°C in an open silica crucible. The melts are very fluid and can quickly be made homogeneous.

In all the transition metal oxide glasses the semi-conductivity is due to the non-stoichiometry of the transition metal oxide. Thus in the vanadate glasses, although most of the vanadium is present as V^{5+} ions, there is a significant proportion of V^{4+} ions also. Conductivity is due to the transfer of electrons between adjacent vanadium ions of different valency. The vanadium ions may be thought of as sites on which an electron is momentarily at rest, a V^{5+} ion representing an empty site and a V^{4+} ion one which is occupied.

The conductivity depends upon the concentration of the transition metal oxide in the glass and on the relative proportion of ions in the higher and lower valence states. The latter is affected by the nature and proportions of other oxides in the glass composition and by the melting conditions, i.e. by the melting temperature, the surrounding atmosphere and the rate of cooling.

In addition to those oxide glasses which are semi-conducting in bulk, there are some glasses which can be treated to produce a semi-conducting surface. Thus glasses containing a high percentage of PbO develop a semi-conducting surface when heated in a reducing atmosphere. This is due to the reduction of the PbO to metallic lead which forms a high concentration of minute particles dispersed in an insulating matrix. This type of treatment is used in the manufacture of a certain type of image intensifier (Chapter VI).

All the semi-conducting glasses are strong absorbers of visible radiation and appear black or show a metallic reflection. Very thin films of some compositions show some transparency and are strongly coloured. Careful examination is necessary to ensure that a glass of this type is free from crystallization and phase separation.

The conductivity of semi-conducting glasses is much less sensitive to impurities than that of crystalline semi-conductors such as silicon and germanium. The explanation for this difference has been given by Mott (1969, 1977). He points out that the random structure of amorphous semi-conductors allows each electron to be used in chemical bond formation. Thus in amorphous germanium, for example, atoms such as phosphorous or antimony, which act as electron donors when added to crystalline germanium, can find sites which are surrounded by five germanium atoms. Thus each of the five valency electrons of the group V impurity element may be used in chemical bond formation. Similarly, group III impurities such as boron which act as electron acceptors in the crystal can find sites in the amorphous material in which the boron atom can bond to three germanium neighbours.

A second noteworthy feature of the semi-conducting glasses of simple composition is that the electrical resistivity of the glass is significantly higher than that of the crystalline form of the same composition. This is illustrated for a number of chalcogenide compounds in Table XXIV. It would seem reasonable

TABLE XXIV

Comparison of Conductivities of Various Chalcogenide
Materials in Their Crystalline and Vitreous Forms

Composition	Conductivity, ohm^{-1} cm^{-1}	
	Glass	Crystal
$2As_2Se_3 . 3As_2Te_3$	2.5×10^{-8}	10^2
$As_2Se_3 . 2Tl_2Se$	7.5×10^{-8}	7.2×10^{-3}
$2As_2Se_3 . Sb_2Se_3$	1.0×10^{-10}	2.5×10^{-6}
$Tl_2Te . As_2Te_3$	5.3×10^{-3}	4.6×10^{-1}
$Tl_2Se . (5/6 \ As \ 1/6 \ Sb)_2Se_3$	2.2×10^{-7}	2.0×10^{-8}

to attribute this effect simply to the greater scattering of electrons which might be expected to occur in the non-periodic structure of the glass. However the understanding of the electrical properties of semi-conducting glasses is difficult and simple qualitative ideas such as this are of little value.

2. Electron Energy Levels and Conduction Mechanisms in Semi-conducting Glasses

It is not possible in a book of this kind to do more than give a brief sketch of the approach which has been made to the study of conduction mechanisms in semi-conducting glasses. This topic is at the forefront of developments in solid state physics and it was for work in this field that Nobel prizes in physics were awarded in 1977 to N.F. Mott and P.W. Anderson. A review article by Owen (1970) discusses the more important features of the experimental results and describes in a relatively simple way the theory of electronic energy levels in non-crystalline solids. A fuller treatment is to be found in the book by Mott and Davis (1971).

The existence of bonds of allowed energy levels in a crystalline solid can be demonstrated by solving the one-dimensional Schrödinger equation

$$\partial^2\psi/\partial x^2 + (8\pi^2 m/h^2)(E-V)\psi = 0 \qquad (103)$$

for a potential V which varies periodically with x, i.e. for a one dimensional crystal consisting of a row of atoms spaced the same distance, a, apart (Fig. 146a). The problem can be solved for a variation of V with x of the form shown in Fig. 146b. This leads to what is called the Krönig-Penny model. However the mathematics is made considerably easier if the heights of the energy barriers, V, are made very large and their widths, b, very small, keeping the value of the product Vb finite (Fig. 146c). The details of the analysis are given in various text books of solid state physics, e.g. Smith (1961). The result of the analysis shows that an electron in a periodic lattice can only have energy values which lie within certain bands. For energies between these bands, solutions of the Shrödinger equation do not exist.

Other factors determine whether the bands of allowed energies contain electrons or not. In an insulator the highest en-

258

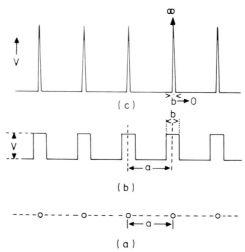

Fig. 146. (a) a regular (crystalline) one-dimensional array
of atoms, (b) potentials of atoms in 'a' approximated by rectangular barriers of width b and height V, (c) the potential
barriers approximated further by δ barriers with b → 0 and
V → ∞. (Owen, 1970).

ergy band containing the valency electrons is completely filled
and the next higher band is empty. The gap, E_g, between the top
of the valence band and the bottom of the next higher band is
typically several electron volts so that the probability of an
electron being excited by thermal energy into the empty band is
very low.   This excitation is necessary for electronic conduction to occur.   In an intrinsic semi-conductor, E_g may be of
the order of 1eV or less.   Electrons are thermally activated
into the empty band even at room temperature and electronic conductivity is observed.   For such a material the variation of
conductivity, σ, with temperature is given by the equation

$$\sigma = \text{const. } \exp(-E_g/2kT) \tag{104}$$

Owen describes the way in which the solution of the simplified Krönig-Penny model is modified when the δ-function energy barriers are not uniformly spaced.   This corresponds to
a one dimensional amorphous solid or glass.   Bands of allowed
energy may still exist, within which an electron can propagate
freely through the material, but in what corresponds to the gap
of forbidden energy levels there is a distribution of allowed
levels which decrease in density as one moves into the gap from

above and below.   This is what is implied by the shaded area in
the density of states diagram in Fig. 147a.   Further considera-

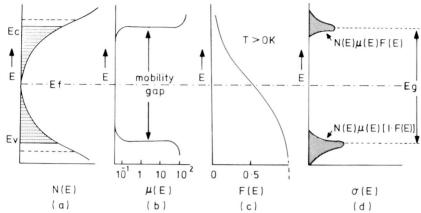

Fig. 147.   (a) Density of states in conduction and valence
bands, (b) the mobility-energy relationship and the 'mobility
gap', (c) the Fermi function for T → 0°K, (d) the contributions to the conductivity from states of different energies.
(Owen, 1970).

tions lead to the conclusion that the electrons which occupy
these tails into the gap are localized at or near certain positions in the disordered structure.   They cannot propagate freely
through the material.   To take part in the conduction process
an electron in a localized state must either be thermally activated into the empty band or must hop directly into a neighbouring localized state.   It has been shown that the electrons in
the localized states have low mobilities.

These complications in the electron energy level scheme in
non-crystalline semi-conductors make it more difficult to interpret the experimental results.   Although some of the semi-conducting glasses, e.g. the chalcogenides and vitreous selenium,
show a simple conductivity temperature relationship which can be
represented by

$$\sigma = const. \exp(- E/kT) , \qquad (105)$$

as may be seen from Fig. 148, E cannot now be equated to $2E_g$.
Because of the low mobility of the localized states, E_g is considered to represent a mobility gap rather than an energy gap.

The semi-conducting oxide glasses such as the vanadate
glasses show a more complicated conductivity relationship. E in

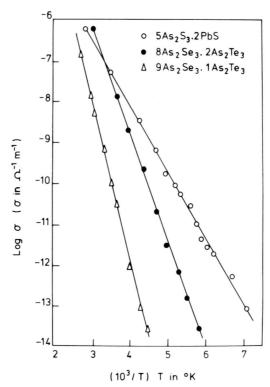

Fig. 148.   Variation with temperature of the conductivity
of three chalcogenide glasses.   (Owen, 1970).

equation 105 varies considerably with temperature (Fig. 149).
As explained in the previous section, the conduction mechanism
in these materials involves the interchange of electrons between
V^{5+} and V^{4+} ions.   The electrons are believed to occupy a range
of localised energy states in the gap, the magnitude of this
range being represented by the symbol W_D.   Conduction involves
hopping directly from one localized state to a neighbouring one.
This should give rise to an activation energy equal to W_D.   However there is likely to be another and larger contribution to
the activation energy arising from the fact that the difference
between the charges of the V^{5+} and V^{4+} ions causes differences
between the vanadium-oxygen distances in the two types of site.
This is illustrated in Fig. 150.   In Fig. 150a the two adjacent vanadium ions are both pentavalent.   For each site the energy of interaction between the vanadium and the surrounding

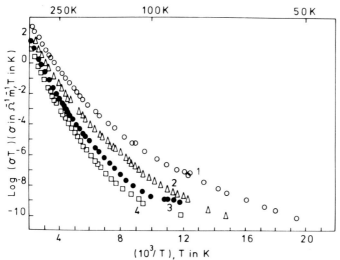

Fig. 149. Variation with temperature of the conductivity of four vanadium phosphate glasses. (Owen, 1970).

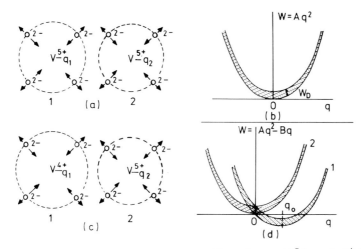

Fig. 150. (a) and (c) represent vanadium ions (V^{5+} and V^{4+}) and their surrounding oxygen ions; (b) and (d) illustrate respectively their energies as a function of the configurational parameter q - a measure of the vanadium-oxygen interionic distance. In (c) and (d) the group labelled 1 is 'polarized' by an extra electron. (Owen, 1970).

oxygens varies parabolically with q, the vanadium-oxygen distance. This variation is similar for the two sites (Fig. 150b). However in Fig. 150c one vanadium site has acquired an extra electron making it a V^{4+} ion. This causes a local polarization of the structure and an increase in the vanadium oxygen distance. The corresponding energy diagram for the site is shown as curve 1 in Fig. 150d. To transfer from a V^{4+} site to a V^{5+} site the electron must acquire an extra energy related to the amount by which the energy of the V^{4+} site has been reduced by the presence of the electron and its associated polarizing effect. The entity of the conduction electron and its associated polarization cloud is referred to as a 'polaron'. There are reasons for believing that both the polarization energy W_p and the energy range of the localized energy levels W_D decrease with decreasing temperature thus giving rise to the variation in activation energy with temperature seen in Fig. 149.

D. Dielectric Loss in Oxide Glasses

1. The Loss Mechanism

The application of an electric field across an insulating material results in a displacement of positive charges relative to negative charges and the material acquires a dipole moment, the magnitude of which is proportional to the field strength. In any particular material the polarization usually involves several kinds of charge displacement. There is some motion of the electrons in individual atoms relative to the nuclei and in ionic crystals there is a small displacement of cations relative to the anions. Motions of this kind occur almost instantaneously as soon as the field is applied. If the field alternates, the polarization remains in phase with the field up to very high frequencies, 10^{12} Hz or even higher. In oxide glasses and ionic crystals, polarization also involves movements of ions over distances of one or two ionic diameters from one equilibrium position to another. These require the surmounting of energy barriers as in the process of d.c. conduction. It takes a finite time for all the ions capable of this kind of motion to reach their final positions after the field is applied. Consequently the polarization varies with time as shown in Fig. 151. In the simplest model which can be used to describe this type of behaviour, the variation of the polarization, P, with time is exponential after the initial instantaneous polarization, i.e.

$$P_t = (P_S - P_\infty) \cdot (1 - \exp((t-t_0)/\tau)) \qquad (106)$$

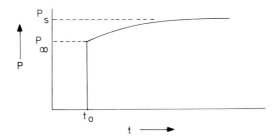

Fig. 151. Time-dependence of polarization after applying
a constant d.c. field at time t_O.

where t_O is the time at which the field is applied and the parameter τ is called the relaxation time. As the temperature of
the material is increased, more thermal energy is available and
the rate at which ions surmount energy barriers to reach their
final position is increased. Consequently τ decreases with increasing temperatures.

If the field alternates, there may be insufficient time in
one half cycle for the ions to reach their final positions and
for the polarization to reach the final value P_S. At very high
frequencies none of the ions may move in one half cycle and the
polarization will then not exceed the instantaneous value, P_∞.

When an ion of unit charge is moved over an energy barrier
through a distance d by a field ξ, the net work done is $\xi e d$
(section B2). A continuous movement backwards and forwards under the action of an alternating field results in a continuous
absorption of energy by the material. The work done on the ions
is dissipated as heat. The resultant rise in temperature decreases the relaxation time, thus increasing the number of ions
which move in one half cycle. This can result in a run-away
situation of progressively increasing energy loss. Obviously
this is a situation to be avoided in a material being used as a
dielectric in a capacitor. On the other hand it is the effect
which makes dielectric heating possible.

The preceding discussion should make it clear that the polarization and the energy loss per cycle vary with frequency.
The energy loss per cycle of material is proportional to the tan-
gent of the loss angle, δ. For an ideal dielectric, δ is zero
and with a condenser made from such a dielectric in an a.c. cir-

cuit the current phasor leads the phasor of the voltage across
the condenser by 90°.   If the dielectric is not ideal, the phase
difference is reduced by δ.   The material property which determines the polarization P for a given applied field E is the dielectric constant ϵ, the three quantities being related by the
equation

$$P = ((\epsilon-1)/4\pi)E \tag{107}$$

The variation of ϵ and tan δ with frequency, ω, for a material
with a single relaxation time τ is given by the Debye equations

$$\epsilon_\omega = \epsilon_\infty + (\epsilon_S-\epsilon_\infty)/(1+\omega^2\tau^2)$$

$$\tan \delta = (\epsilon_S-\epsilon_\infty)\omega\tau/(\epsilon_S+\epsilon_\omega\omega^2\tau^2) \tag{108}$$

ϵ_S and ϵ_ω are respectively the limiting values of the low frequency and high frequency dielectric constants, corresponding
to P_S and P_∞

Figure 152 shows the variation with frequency of

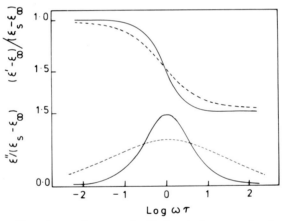

Fig. 152.   The dependence of dielectric constant and loss
(both normalised) on frequency.   (Owen, 1963).

$(\epsilon_\omega-\epsilon_\infty)/(\epsilon_S-\epsilon_\infty)$ and ϵ_ω tan $\delta/(\epsilon_S-\epsilon_\infty)$.   The solid lines are those
given by the Debye equations, i.e. they apply to a material with
a single relaxation time.   The value of ϵ_ω tan δ has its maximum value at the frequency $\omega = 1/\tau$ whilst tan δ reaches its maximum value at a slightly different frequency $\tau^{-1}(\epsilon_S/\epsilon_\infty)^{\frac{1}{2}}$.

As might be expected, the dielectric behaviour of a glass
is too complicated to be represented by a single relaxation time.
The dotted lines in Fig. 152 represent the a.c. behaviour of a

typical glass.   Clearly a distribution of relaxation times is
involved.

Since the d.c. conductivity of an ionically conducting
glass involves the thermally activated motion of alkali ions,
one might expect to be able to calculate the dielectric loss
from the measured electrical resistivity.   However there are
clearly differences between the mechanisms of d.c. conduction
and a.c. loss.   The former involves a steady motion of ions
through the material for as long as the field is applied whilst
the latter involves a more limited motion which decays with time.

It is true however that the ion movement associated with
the d.c. conduction process does contribute to the a.c. loss,
especially at low frequencies.   The a.c. behaviour of a dielectric which is not a perfect insulator can be modelled by the
behaviour of a condensor and resistor in parallel.   Using elementary a.c. circuit theory to work out the loss angle of the
circuit gives $\tan \delta = 1/\omega RC$.   Replacing component values in the
equation by material properties, $\tan \delta = 1/\omega \rho \epsilon$.   Inserting typical values of ρ and ϵ, the value of $\tan \delta$ at 1 kHz is found to
be of the order 10^{-4} at room temperature, which is about 100
times smaller than the measured value for a soda-lime-silica
glass.   However at elevated temperatures and at lower frequencies the d.c. resistance component makes a significant contribution to the loss.

It is clear therefore that the time-dependent dielectric
polarization of a glass involves the movement of a far larger
number of alkali ions than is involved in the d.c. conduction.
A physical model which could account qualitatively for the different types of ion movement can be visualized by considering
the energy level diagram in Fig. 136.   Clearly more ions will
have enough thermal energy to surmount the smaller energy barrier than the larger ones.   The a.c. losses in the frequency
range between 1 kHz and 1 MHz can be visualized as involving the
motion of ions backwards and forwards across the lower energy
barriers.   However it will be seen in the next section that
this simple interpretation is not entirely satisfactory.

266

2. The Frequency and Temperature Dependence of the Dielectric Loss of Alkali Silicate Glasses

The most detailed study of the dielectric properties of silicate glasses in the low to medium frequency region has been carried out by Taylor (1957, 1959). The glasses included several compositions in the Na_2O-SiO_2 and $Na_2O-CaO-SiO_2$ system and two commercial glasses. Measurements were made from $100-10^4$ Hz and from room temperature to 300°C.

The measured variation of dielectric loss with frequency was not found to be of the form shown in Fig. 152, with a well defined maximum. However, when the contribution from the d.c. conduction component is subtracted from the measured values, curves of the expected Debye form are obtained (Fig. 153). The

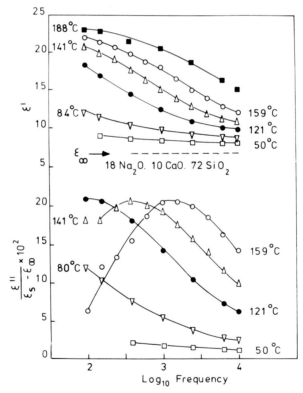

Fig. 153. Dielectric dispersion and absorption curves (corrected for d.c. conductivity) for a soda-lime-silica glass. (Taylor, 1959).

increase with temperature of the frequency at which the loss is a maximum implies a corresponding decrease in the mean relaxation time τ. Since the loss process involves a thermally activated motion of alkali ions, τ is expected to vary with temperature as

$$\tau = \text{constant}.\exp(E_{a.c.}/RT) \tag{109}$$

where $E_{a.c.}$ is the activation energy controlling the motion. Surprisingly, in view of the discussion in the previous section, this turns out to be almost identical with the activation energy for d.c. conduction. Taylor's values for the two activation energies are compared in Table XXV.

TABLE XXV

Comparison of Activation Energies for d.c. Conduction (E_{dc}) and Dielectric Relaxation (E_{ac}) - Taylor (1956)

Glass (Mole %)	E_{ac}(kcal/mole)	E_{dc}(kcal/mole)
$18Na_2O.10CaO.72SiO_2$	19.8	19.3
$10Na_2O.20CaO.70SiO_2$	25.0	23.3
$12Na_2O.88SiO_2$	17.0	16.7
$24Na_2O.76SiO_2$	16.5	16.4

The variation of the dielectric loss of silicate glasses over the whole range of practical interest up to microwave frequencies has been discussed by Stevels (1957). Figure 154 shows this variation schematically. The dotted curves represent what are suggested to be the separate components which contribute to the loss. Two of these have already been discussed. Curve 1 is the contribution from the conduction current and curve 2 that from the ionic relaxation mechanism. Of the other two mechanisms only one, represented by curve 3, is found to be markedly affected by temperature. Stevels attributes this to what he calls a deformation loss mechanism, the deformation involving a relaxation within the silicon-oxygen network. Curve 4 is due to vibration losses similar in kind to those responsible for infra-red absorptions, but occurring at much longer wavelengths.

Apart from Taylor's work, relatively little systematic work has been done on the effects of glass composition. As one might expect, the main compositional factor determining the loss is

268

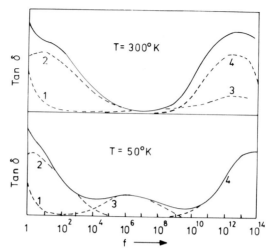

Fig. 154. Schematic representation of the variation of
dielectric loss of an alkali-containing glass with frequency
and temperature. (Stevels, 1957).

the alkali content. Data on commercial glasses has been col-
lected together by von Hippel (1954).

CHAPTER IX

CHEMICAL DURABILITY

A. Introduction

Everyday experience suggests that glasses are relatively inert chemically. Corroded metal is a more common sight than badly weathered glass. Many specimens of ancient Egyptian glass
exist which appear to be as bright and glossy as they probably
were when they were made.

For many applications the chemical durability of glasses is
perfectly adequate. However conditions are occasionally encountered when one becomes aware of the fact that even container and
window glass compositions are not completely resistant to atmospheric attack. If they are exposed to warm, humid air, they become covered with a white crust formed by reaction between the
glass surface and the moisture in the atmosphere. In a stationary atmosphere containing water vapour, the film of water which
forms on the glass surface becomes highly alkaline and very corrosive as sodium ions are leached out from the glass. Faulty
double glazed windows show the effect very clearly (Fig. 155).

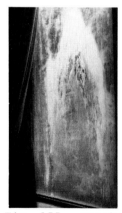

Fig. 155. Weathering of the internal surfaces of a faulty
double glazed window.

These units are hermetically sealed, the air between the panes
being carefully dried. If the seal leaks, letting in atmospheric moisture, the ideal situation for accelerated attack of the
internal glass surfaces soon develops. A window exposed to the
weather is not attacked at anything like the same rate since the

water film on the glass surface is never static for long periods so that there is insufficient time for it to extract enough alkali to become corrosive.

It is possible, but it would be uneconomic, to make windows and glass containers from compositions which are resistant to these unfavourable conditions. However it is more reasonable to avoid them. For example, if glassware is to be stored for long periods, the warehouse should be dry, cool and well-ventilated, and water-absorbing packing materials should not be used.

If the corrosion is so slight that only a faint white bloom forms on the glass, this can sometimes be removed chemically. If the glass cannot be restored to what appears to be its original condition by a water wash, immersing it for a few minutes in a hot solution of a chelating agent such as E.D.T.A. may prove successful or, failing that, a dilute solution of hydrofluoric acid (Ernsberger, 1959; Tichane and Carrier, 1961; Tichane, 1966). However if the attack is so severe as to require the use of hydrofluoric acid, it may be cheaper (and safer) to replace the glass. With irreplaceable items, such as panes of medieval glass, one may be prepared to go to considerable lengths to remove the corroded layer and to protect the restored window from subsequent corrosion. At the present time a considerable amount of effort is being expended in Europe on restoration work, much of the need for this having been brought about by the storage of the glass under unsuitable conditions during the Second World War (Newton, 1982).

Although the attack on glass containers by their contents rarely presents any problems, one has to remember that trace quantities of materials extracted from the glass may have serious effects. There are a number of applications of containers where it is important to avoid contamination, the most important being the use of glass ampoules for containing drug solutions. Care is also necessary in selecting glass enamels which are sometimes used to decorate the rims of drinking glasses. In the past, many of these contained a high percentage of lead oxide. These must now be avoided or, if that is not possible, it is necessary to ensure that the rate of extraction of lead from the enamel is not so high as to expose the user to the risk of cumulative lead poisoning. The risk is greater with ceramic articles such as jugs which may be glazed internally with a lead-containing glaze. If used to contain a slightly acid liquid, e.g. fruit juices, the lead extraction may be considerable. Modern factory-produced ware is unlikely to be defective in this respect, but cases are known where the products of the small craft potter have proved fatal. There is said to be some risk of lead poisoning from drinking whisky which has been kept for a long time in a lead glass decanter. However the risk is probably less than that of drinking the whisky in haste!

Some special glasses, e.g. optical glasses, are less durable than container and flat glass compositions. Although they are formulated primarily for their optical properties, it is always necessary to ensure that they are sufficiently durable to withstand atmospheric attack in whatever part of the world they may be used. This consideration has an important effect in formulating the glass composition.

A particularly critical example of formulating a glass composition which is resistant to the leaching out of dangerous constituents arises in the project for storing radioactive wastes from nuclear reactors in the form of a glass made by fusing the wastes with other oxides (McElroy, 1975; Corbet, 1976; Johnson and Marples, 1979). The dangers to the environment of escape of these wastes is well-known and a great deal of work is being done to ensure that the risks involved in using this method are acceptable.

Finally, there are applications for which glasses have been developed for withstanding especially severe attack, e.g. glass tubes for boiler water-level indicators, sodium-resistant glasses for vapour discharge lamps and glass fibres for reinforcing cement. In the latter application, the glass is exposed to attack by the highly alkaline aqueous solution formed by reaction between water and the cement. Any attack on the fibres results in a loss in strength and the reinforcement effect is impaired.

The literature on the chemical durability of glasses is extensive. A number of valuable review articles and bibliographies exist, e.g. Holland, 1963; I.C.G., 1965, 1972, 1973; Bacon, 1968; Hench, 1977; Paul, 1977. Current knowledge in the field is difficult to summarize. Generally speaking, prac-

tical knowledge of technological value is considerably more extensive than that which helps in understanding the mechanisms of
the reactions involved. This is not surprising. The study of
surface reactions has always been difficult and it is only relatively recently that instruments have become available which allow one to analyze the attacked surface and the very small quantities of material extracted from the glass.

B. Weathering

This term is applied to the process in which a glass surface reacts with a water-vapour containing atmosphere to produce
a visible surface film or crust.

The initial rate of weathering can be followed using optical techniques such as ellipsometry which measure the refractive
index and thickness of the surface film (Tsuchihashi et al.,1975).
This is possible only when the film is of uniform thickness and
is sufficiently thin, i.e. of the order of one wavelength. However, electron microscopic examination of weathering films in
the early stages of the process indicates that attack frequently
begins at nuclei distributed over the surface, from which crystals of the reaction product grow until the surface is eventually covered by a continuous crystalline layer (Tichane, 1966).
The thickness of the layer then grows until the surface is covered with an opaque, white crust.

In practice one is usually most interested in the visual
effect of the attack. This is difficult to measure and express
quantitatively since the rate of attack often varies markedly
over the surface of a test specimen and may even show localized
pitting. One method of measuring the degree of weathering involves the use of a "haze meter" (Simpson, 1951). Flat polished
glass specimens are normally used. After test the percentage
haze, measured by this instrument, is obtained from two light
intensity readings. One is for the intensity of light transmitted through the specimen normal to its surfaces and the other
is the total intensity transmitted, i.e. the sum of the directly
transmitted and the scattered light. According to Walters and
Adams (1975), this technique is not sufficiently sensitive to
detect differences between glasses in the early stages of weathering. They prefer a simple order of merit assessment with a

system of grading based on a visual comparison of the specimens
under test with a series of standards.

When carrying out weathering tests in the laboratory, the
specimens must be subjected to an accelerated attack in order to
obtain results in a reasonably short time.   They are exposed at
an elevated temperature to an atmosphere of relatively high humidity.   The apparatus used by Simpson (1951) is shown in Fig.
156.   The specimens were supported on a 'Pyrex' glass stand over

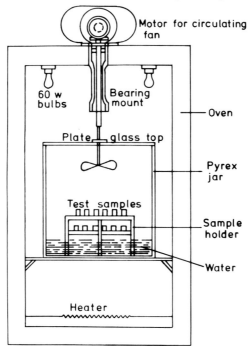

Fig. 156.   Apparatus for accelerated weathering tests
(Simpson, 1951).

water contained in a large 'Pyrex' jar.   The jar was inside a
modified oven, the temperature of which could be cycled from 50°
to 55° over a two hour period.   This caused water to alternately
condense on and evaporate from the surface of the specimens.
It is generally believed that under service conditions this alternate condensation and evaporation increases the rate of attack.
This view is not supported by the recent work of Walters and
Adams (loc. cit.) who found in their accelerated tests that constant conditions of humidity and temperature usually resulted in

the more severe attack.

The degree of visible attack is greatly reduced if the weathering products are washed off the surface (Fig. 157), an

Fig. 157. Comparison of washed and unwashed soda-lime glass tumblers after weathering at 98% relative humidity and 50°C. (Walters and Adams, 1975).

observation which suggests that the layer of weathering products does not act as a barrier layer reducing the rate of attack between the glass and the atmosphere as some oxide films do on metals.

Figure 158 shows some of the weathering results obtained by Walters and Adams on a number of commercial silicate glasses. Class A signifies that no spots or haze can be seen when the specimens are illuminated by a concentrated beam of light, whilst class E signifies a large accumulation of weathering products, easily visible under normal daylight conditions. It is worth noting that there are marked differences between the rates of weathering of the various types of commercial soda-lime-silica glasses, even though the differences between the compositions are small. This is in agreement with the earlier work of Simpson (1959). It is not easy to relate the resistance to weathering, assessed by its visual effect, to the glass composition. However generally speaking one can say that the weathering resistance of the soda-lime-silica types of glass is considerably improved by relatively small reductions in alkali content and small increases in alumina content.

It is generally agreed that the weathering process involves

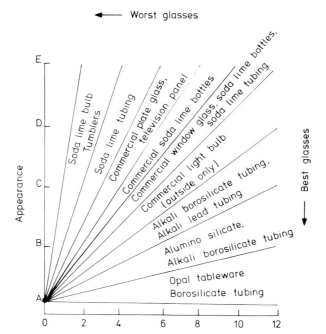

Fig. 158. Average weatherability of various glasses under high humidity (Walters and Adams, 1975).

the adsorption of water onto the glass surface, followed by a base-exchange reaction between hydrogen ions supplied by the water and sodium ions in the glass. The highly alkaline solution so formed then begins to break down the silica network. Crystalline products of complicated composition may subsequently precipitate from the solution onto the glass surface. Walters and Adams measured the rate of extraction of alkali from the glass during their weathering tests. This was done by washing the surfaces with distilled water and measuring the amount of alkali in the washings using a flame photometer. Relatively little work of this kind has been done. For the least durable of the commercial soda-lime-silica glasses, the amount of alkali extracted was more than 5 µg per square centimetre of specimen surface after exposure for 14 days at 50°C to air of 98% relative humidity. This corresponds to the total alkali content of a layer of glass 200 nm thick. Although this is less than half a wavelength, the visible effect of the alkali was quite pronounced.

C. Reactions of Silicate Glasses with Aqueous Solutions

1. Methods of Measurement

Many studies were made during the last century of the reactions of silicate glasses with aqueous solutions and it is more
than fifty years since the first systematic studies were made of
the effects of changes in glass composition on the reaction rates.
This early work was clearly necessary for the development of durable container glasses and, no doubt, practical considerations
motivated much of it. During the last ten or fifteen years,
more detailed work has been done in an attempt to elucidate the
reaction mechanisms. Although this has yielded interesting and
useful results, it has also shown that the reactions are complicated and may change in nature during the course of an experiment.
When faced with such a complicated situation, one may have to
consider whether the scientific and technological value of much
further basic work in this field is likely to justify the considerable effort which is clearly needed. However there are
many good reasons for needing to know more about the structure
and composition of glass surfaces, so no doubt basic work will
continue even though progress is likely to be slow.

The different constituents are extracted from the glass at
different rates resulting in the formation of a surface layer
which has a composition different from that of the bulk glass.
The thickness of the layer increases with time and this is one
of the factors that can cause a change in the nature of the reaction.

To study the reaction in detail it is necessary, amongst
other things, to follow the increase with time of the various
glass constituents in the solution. This is made easier by
crushing the glass into granules of carefully controlled grain
size thus providing a large surface for attack and/or by working
on simple binary or ternary compositions which are usually more
rapidly attacked than the more complex commercial compositions.

Two procedures have been adopted to reduce the changes in
the conditions of attack during an experiment which result from
the change in composition of the aqueous solution as it takes up
glass components. In one the glass does not come into contact
with the aqueous solution itself (Fig. 159). The glass gran-

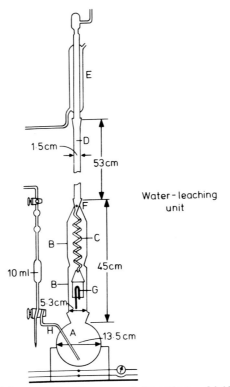

Fig. 159. Water-leaching unit (Rana and Douglas, 1961).

ules are contained in a vessel, G, which is suspended above the
solution in the neck of a silver flask, A.   The solution in the
flask is boiled at the pressure within the apparatus, which is
automatically controlled.   Water vapour escapes into the water
condenser, D, the condensed water flowing back through the spiral,
C, onto the glass in G.   When this is full, the water syphons
back into the flask.   Samples of the solution are taken periodically from the flask for analysis.   By controlling the pressure
in the apparatus, experiments can be conveniently carried out at
any temperature in the range from room temperature to 100°C
(Beattie, 1953;   Rana and Douglas, 1961).

In the second method, the glass granules are brought into
direct contact with the aqueous solution but, by using buffered
solutions, changes in their pH are greatly reduced (El-Shamy et
al., 1972).

2. Effect of pH of the Aqueous Solution

This has been investigated and discussed by El-Shamy and Douglas (1967) and El-Shamy et al. (1972). For vitreous silica it was found that the rate of attack was extremely low up to a pH of 9, but more alkaline solutions attacked the glass much more rapidly (Fig. 160). In all the experiments on vitreous silica

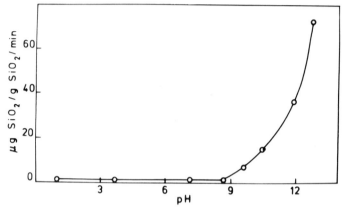

Fig. 160. Effect of pH on the rate of silica extraction from vitreous silica powder at 80°C (El-Shamy and Douglas,1967)

the rate of increase of the silica content of the buffer solution was constant over the period of the experiment (120 minutes).

Similar sudden changes with pH in the rates of reaction were observed for binary and ternary silicate glasses. Also for many of them it was found that the quantity, Q, of alkali extracted in time t varied linearly with $t^{\frac{1}{2}}$, at least in the early stages of the reaction. Hence the product $Q.t^{-\frac{1}{2}}$ can be used as a measure of the rate constant, k, of the reaction. The variation of k with pH is shown in Fig. 161 for a glass of composition 15 mol% Na_2O, 85% SiO_2. The marked change in reaction rate is again found at pH 9. Beyond that point the rate of extraction of sodium decreases as the solution becomes more alkaline. However the rate of extraction of silica shows an opposite trend, which is illustrated in Fig. 162 for a glass of composition 15 mole% K_2O, 85% SiO_2. The quantities of silica extracted were too small to determine how the extent of reaction depended on time. Consequently the results are expressed as the quantity of silica extracted over a specified time period.

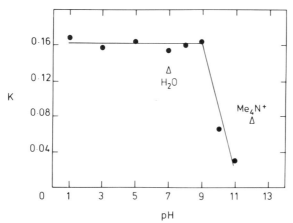

Fig. 161. Dependence on pH of the rate of extraction of sodium oxide (in mg g^{-1} min$^{-\frac{1}{2}}$) from a glass of composition 15 mol% Na$_2$O, 85% SiO$_2$ at 35°C (El-Shamy et al., 1972).

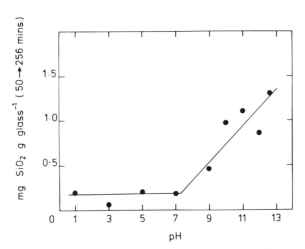

Fig. 162. Dependence on pH of the rate of silica extraction at 35°C from a glass of composition 15 mol% K$_2$O, 85% SiO$_2$ (El-Shamy and Douglas, 1967).

Figure 163 shows very well the importance of controlling the pH of the attacking solution. This shows the rate of extraction of silica by water from granules of a sodium silicate glass. The difference between the two sets of results arises from the fact that for the results given by the line B a cation exchange resin was added to the water. Alkali ions brought in-

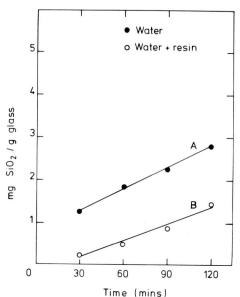

Fig. 163. Effect of removal of alkali from attacking
solution on silica extraction (El-Shamy and Douglas, 1967).

to the solution as a result of base exchange are believed to be
responsible for the breakdown of the silica network of the glass.
The resin removes these ions from the solution, thus causing a
reduction in the rate of attack.

In the discussion of the reactions in the papers referred
to earlier, it is suggested that the first stage involves the
formation of an alkali-deficient surface layer on the glass as a
result of the more rapid extraction of alkali. Subsequent extraction of alkali involves diffusion of alkali through the layer from the glass towards the solution. Hydrogen ions from the
solution diffuse in the opposite direction to maintain charge
balance. The rate of extraction of alkali is proportional to
the concentration gradient in the reaction layer, giving rise to
the $t^{\frac{1}{2}}$ dependence of the quantity of alkali extracted. It is
suggested that at values of pH greater than 9, surface sites at
the interface between the layer and the solution become increasingly occupied by alkali ions, thus reducing the concentration
gradient in the film and hence the rate of reaction. Opposing
this effect is the increasing rate of removal of silica into the
solution with increasing pH, which is likely to have the effect

of reducing the thickness of the silica surface layer.

3. Reaction with "Pure" Water

The method in which the composition of the aqueous phase is
kept reasonably constant by bringing the glass into contact with
a continuous flow of distilled water has been used by Beattie
(1953), Rana and Douglas (1961) and Das and Douglas (1967).
In the early stages of the reaction, the quantity, Q, of each
constituent extracted increases linearly as $t^{\frac{1}{2}}$ but at long times
the rates of extraction become linear, i.e. Q increases linearly
with t.   This is illustrated in Figs. 164 and 165 by the results

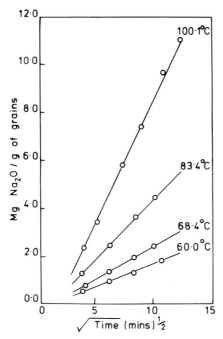

Fig. 164.   Kinetics of the extraction of alkali from a glass
of composition 20 mol% Na_2O, 80% SiO_2 in the early stage of
the reaction (Rana and Douglas, 1961).

for a glass of composition 20 mol% Na_2O, 80% SiO_2. The explanation offered for the change in the reaction kinetics is that in
the early stages ($t^{\frac{1}{2}}$ dependence) the reaction is a diffusion controlled base-exchange involving sodium and hydrogen moving in
opposite directions through a siliceous surface film, whilst
later (t dependence) the surface film has attained a constant
thickness. This thickness is determined by a balance between

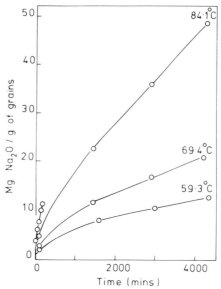

Fig. 165. Kinetics of the extraction of alkali from a glass
of composition 20 mol% Na_2O, 80% SiO_2 showing the eventual
establishment of a linear time dependence (Rana and Douglas,
1961).

the diffusion of ions through the film, which thickens it, and
the rate of removal of silica at the film-solution interface
which has the opposite effect.

For a commercial soda-lime-silica glass, Isard and Douglas
(1949) were able to obtain a quantitative correlation between
the rate of removal of alkali by water attack and the electrical
conductivity, both processes involving the diffusion of sodium
ions. However the results of Das and Douglas (loc cit.) show
that the correlation does not exist for less durable glasses,
the rate of alkali extraction corresponding to values for the
alkali ion diffusion coefficient up to 10^4 to 10^5 times greater
than that obtained from conductivity measurements.

The replacement of silica by a third oxide has the effect
of greatly reducing the rate of attack as shown, for example, by
the results in Fig. 166 for the extraction of alkali from glasses
of the general composition 15 mol% Na_2O, x% R_mO_n, (85-x)% SiO_2.
Das and Douglas studied 12 glasses of the general composition
15 mol% Na_2O, 5% R_mO_n, 80% SiO_2 and showed that the glasses containing TiO_2 or ZrO_2 were especially durable.

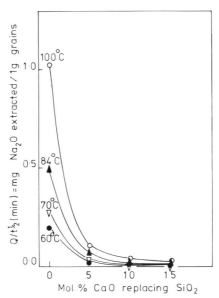

Fig. 166. Effect of replacing SiO$_2$ by CaO on the durability. (Rana and Douglas, 1961).

Generally speaking, the rate of extraction of alkali is greater than the rate of extraction of silica even in the experiments using buffer solutions of high pH; but for the more durable glasses studied by Rana and Douglas the SiO$_2$/alkali ratio in the solution was almost equal to that in the glass. When these glasses were tested in the form of rods, their surfaces after attack were still smooth and shiny. In contrast, the thick siliceous layers which formed on the less durable glasses were translucent in appearance and formed a white friable crust when the specimens were dried. When rods of the least durable potassium silicate glasses were tested, practically all the potassium was eventually extracted, leaving behind a fragile skeleton of silica.

4. The Durability of Three-Component Silicate Glasses and Commercial Compositions

The investigation of the durability of three-component silicate glasses referred to in the previous section was carried out in the context of a study of reaction mechanisms and the compositions were chosen to have the same molecular percentage of Na$_2$O to provide a reasonable basis for the comparison of the effects

284

of a third oxide component. Much earlier work has been carried
out on three-component glasses without the choice of composition
being influenced in this way. However the results are generally
similar to those of the more recent investigations. Figure 167

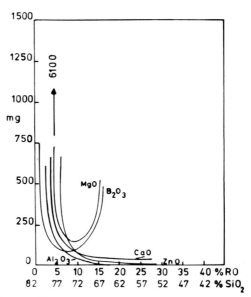

Fig. 167. Effect of replacing SiO$_2$ by various oxides on
the durability (Enss, 1928).

shows the results of Enss (1928) for glasses in the system 18
weight% Na$_2$O, x% R$_2$O, (82-x)% SiO$_2$. He measured the weight of
material extracted from glass granules after heating for 7 hours
in contact with boiling water. The weight extracted from the
binary glass was 6100 mg. Thus the substitution of as little
as 5% Al$_2$O$_3$ for silica in the glass reduces the attack by a factor as much as 60.

The high durability of ZrO$_2$-containing glasses, especially
to attack by alkaline solutions, has recently found an interesting and important application. Fibres of high ZrO$_2$ content
glass (16 weight%) are now being widely used for the reinforcement of Portland cement. The early work at the U.K. Building
Research Establishment (Larner et al., 1976) has been commercially exploited by Pilkington Brothers, the fibre product being
marketed under the name "Cemfil". The glass composition is very
similar to an alkali-resistant glass which has been marketed for

some time in bulk by Corning Glass Limited under the code number 7280. The composition of the latter is SiO_2 71.3 wt.%, ZrO_2 15.8%, Na_2O 11.5%, Li_2O 0.8%, K_2O 1%, MgO 0.1%, CaO 0.1% (Bacon, 1968).

Clark et al., (1976) have shown that CaO is present, no doubt in combined form, in the reaction layer on the surface of CaO-containing glasses that have been attacked by aqueous solutions. They suggest that the CaO has "a stabilizing effect" on the layer and reduces the rate of diffusion of sodium ions through it. Although the improvements caused by the addition of Al_2O_3 and ZrO_2 may have a similar explanation, Paul (1977) suggests an alternative explanation based on thermodynamic rather than kinetic factors. He draws attention to data which indicate that ions containing aluminium and zirconium have a very low solubility in aqueous solutions. Hence equilibrium concentrations of these ions are soon built up in the solution before the reaction with the glass has proceeded very far.

For laboratory chemical apparatus and in the construction of glass chemical plant where resistance to neutral or strong acid solutions is required, the glass composition most commonly used is a borosilicate containing approximately 12 wt.% B_2O_3. One glass of this type is Corning 7740, the composition of which is SiO_2 80.3 wt.%, B_2O_3 13%, Na_2O 4.2%, Al_2O_3 2.4% (Bacon, 1968). The introduction of B_2O_3 makes it possible to melt such a high silica, low alkali composition at temperatures only 100°C or so above those used for making container and flat glass compositions which contain three to four times as much alkali. Borosilicate glasses are also used to make ampoules for containing drug solutions. A typical composition is that of Kimble N51A: SiO_2 74.4 wt.%, B_2O_3 9.5%, Al_2O_3 5.5%, BaO 2.2%, CaO 0.9%, MgO 0.3%, Na_2O 6.6%, K_2O 0.6% (Bacon, 1968).

When developing borosilicate glasses for chemical ware, one needs to guard against the tendency of some alkali borosilicate compositions to undergo phase separation on heat treatment below the liquidus temperature. One phase is of high alkali content and is readily attacked by aqueous solutions. The large effect of heat treatment on the chemical durability of a borosilicate glass is clearly shown in the work of Howell et al., (1975).

A useful experimental survey by Cameron and Horne (1964) of

the durability of a wide range of commercial glasses under attack by dilute acid and alkaline solutions is worth mentioning
for reference purposes.

5. The Effect of Temperature on the Rate of Attack by
    Aqueous Solutions

Investigations of the effect of temperature on the rate of
attack of silicate glasses by water, steam, dilute acids and dilute alkaline solutions, have been summarized by Holland (1963),
Most of the results fit an Arrhenius equation

$$S = const.exp(- E/RT)$$

where S is the rate of attack and E is an activation energy.
The values of E usually fall in the range between 15 and 19 K
cal mole^{-1}, i.e. approximately equal to that for the diffusion
of sodium ions in a silicate glass.   Although this is not surprising in view of the earlier discussion suggesting that one of
the reactions involves base-exchange through the surface layer
of sodium and hydrogen ions, one might expect that more detailed
work on a limited range of glass compositions using solutions
over a wide range of pH values would show a variety of temperature effects as other reaction mechanisms are brought into effect by changes in the experimental conditions.

D. Standard Tests for Chemical Durability

It should be clear from the previous section that it is
difficult, if not impossible, to make general statements about
the durability of one glass compared with another.   Not only do
the results depend on such obvious factors as temperature and
the composition of the aqueous phase, they are also affected by
relatively small differences in specimen preparation, experimental technique and the design of the apparatus used.   It is very
difficult therefore to propose standard tests which are likely
to be generally acceptable, especially if the test is intended
to provide information relevant to reaction mechanisms.   However
this is not the context within which these standard tests are
proposed.   One type of test is to provide a reasonably convenient procedure for making comparisons between commercial glass
compositions, and the second main type is for ensuring that glassware made for particular compositions is satisfactory for its

intended application.

For comparing glass compositions, tests involving the use
of glass granules are the most generally favoured. Apart from
increasing the surface area under attack, thus simplifying the
analysis of the extract, the use of granules eliminates what
can be quite large differences when bulk specimens are tested.
Results on a bulk specimen can be affected by the way in which
its surface was treated before the test. The composition of a
glass surface which has been rapidly cooled by casting the melt
into a metal mould may be quite different from that of a specimen which has been more slowly cooled in contact with the air.
For glasses of low durability, exposure of the specimen to the
atmosphere before the test is likely to affect the results.
Mechanical preparation of the surface by grinding and polishing
does not get round the difficulty since this too may chemically
modify the surface. However Hench (1977) considers that the
use of bulk specimens which have been given a controlled surface
grinding treatment is in fact preferable to the use of grain
tests. Bulk specimens, as he rightly points out, have the very
considerable advantage for reaction mechanism studies that it is
possible to analyse the composition of the attacked surface.
However, the only standard tests at the moment are those involving the use of glass granules. The need to standardize the
temperature, the weight of the sample, and the composition of
the aqueous solution have been well taken care of in the test
specifications for some time. More recently Sykes (1965) has
drawn attention to the need for more care in crushing and siev-
ing the sample to ensure a reproducible surface area per unit
volume.

In testing the durability of glassware, the test is designed so far as possible to simulate the conditions to which the
glass is subjected in service. However to provide results within a reasonable time, the temperature of the test is well above
that encountered in practice. There is always a danger inherent in accelerated tests. The difference between test and service conditions can result in the test placing materials in a
different order of merit from that found in normal service. This
problem is not often of great concern in the durability testing
of glassware. More often than not the tests are being carried

out on ware made from well-established compositions to satisfy a customer who will use the glassware in a well-established application.   It is then simply a question of ensuring that a particular batch of ware is within specification.

Various national standards exist for testing glass and glassware.   Bacon (1968) summarizes the three American standard tests described in detail in A.S.T.M. specification C225-65. More recent developments of standard grain tests are described in various I.S.O. standards (I.S.O. 1968a, I.S.O. 1968b, I.S.O. 1970).   Although the standards specify the test procedures, they do not define the maximum quantity of material that may be extracted from the glass for particular applications.   This may be left to negotiation between the glass manufacturer and the customer.   However for critical applications, e.g. for drug ampoules, performance standards are normally laid down by independent bodies (European Pharmacopoeia, 1971;   I.S.O., 1977).

E.   Methods of Improving the Durability of Glassware or Reducing the Attack by Specific Reagents

Since the attack on glass of commercial soda-lime-silica types involves the extraction of alkali from the surface, it is not surprising that various kinds of surface treatment of glassware which remove alkali from the surface have been found greatly to improve the durability of the ware.

The treatment with the longest history involves exposing the hot glass to air containing a small percentage of sulphur dioxide.   This treatment was applied unknowingly many years ago when annealing lehrs were heated by coal, and later by gas which contained sulphur.   During annealing, the ware became covered with a white deposit of sodium sulphate which was washed off when the glass was cold.

Early laboratory studies of the treatment and its results were made by Cousen and Peddle (1936, 1937).   More recent papers by Isard and Douglas (1949) and Mochel et al., (1966) deal with the mechanism of the surface de-alkalization in more detail. Persson (1962) also included this method in a comparison of various alternative processes, which also included exposing the glass to the vapours of $AlCl_3$ and NH_4Cl.   These remove the alkali to form a film of sodium chloride on the surface, which, like the sulphate, may be removed by washing.

It is clear from the more recent investigations that the presence of oxygen and water vapour in the atmosphere considerably increases the rate of extraction of alkali in the SO_2 process. Isard and Douglas (1949) suggest that the formation of the low alkali surface layer involves a base-exchange between hydrogen ions supplied from the atmosphere and sodium ions from the glass. They also believe that, either concurrently or subsequently, the hydrogen ions are expelled from the surface layers as H_2O. Persson's results show that the alkali content of the glass surface is reduced to a depth of the order of several thousand A.U.

It is interesting to note that even in the absence of materials deliberately introduced to remove alkali from the surface, a detectable loss may occur simply as a result of volatilization as the surface of the glassware is formed. The electron microprobe results of Sieger (1975) in Fig. 168 show that the alkali con-

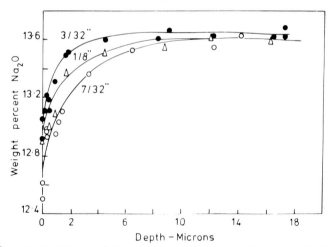

Fig. 168. Variation of Na_2O content near the top surface of float glass (Sieger, 1975).

tent near the surface of float glass is several percent below that of the interior. The results are for the upper surface, i.e. that which was exposed to the bath atmosphere whilst the ribbon of glass was being formed.

Significant surface loss of alkali also occurs in the manufacture of glass containers even though the new glass surface which is produced in making the container is at a high tempera-

ture for only a few seconds. Budd and Kirwan (1962) have measured the "free" alkali present on the internal surface of containers of various designs by quickly rinsing out the cold article with distilled water and analyzing the washings. The quantity of this alkali is much greater, the smaller the diameter of
the container neck. The likely explanation of this effect is
that the alkali vapour can escape more readily from the inside
of a wide necked container; if the neck is narrow, little escapes whilst that remaining condenses on the internal surface
as the container cools. The Auger electron spectroscopy analysis on the internal surface of a container by Pantano et al.,
(1975) shown in Fig· 169 may not be typical but it does con-

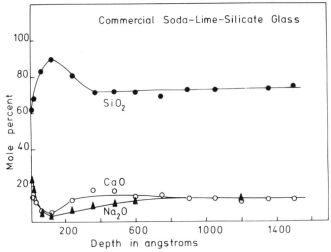

Fig. 169. Results of an Auger electron spectroscopy analysis
of the surface of a glass container (Pantano et al., 1975).

firm loss of alkali to a considerable depth. It is clear from
these observations that containers of different designs made
from glass of the same composition may show significant differences in a chemical durability test and that even for a given
design there may be differences in performance if the conditions
of fabrication change from time to time.

It is interesting to speculate on whether these variations
in surface composition, which have been shown to extend to depths
of 200-300 A.U., may have a significant effect on the viscosity
of the surface layers as they are formed and hence on the way in

which the glass flows during the forming process. It is well-known, especially in the container industry, that periods of poor glass workability, with a high percentage of defective ware, are sometimes encountered, even though conventional chemical analysis and measurements of physical properties suggest that the glass composition is under satisfactory control. It is worth noting in this context the results of work done some years ago by Poole (1967). He attempted to correlate the percentage of defective containers made over a considerable period of time with the results of all the material control measurements made on the glass and on the containers themselves. The only correlation which appeared to be significant was with the alkali extracted in the standard container durability test!

These observations amplify the point made in the previous section that the results of durability tests on bulk specimens of glass may be made more difficult to interpret by variations in alkali lost from the surface of the specimens as they are formed.

To conclude this section, it is worth referring briefly to the fact that the attack on glassware during industrial washing processes may be reduced not only by surface treatment of the glassware, but possibly more economically by including additives in the washing solution. Thus the attack by the 3% sodium hydroxide solution sometimes used in washing milk bottles can be reduced by a factor of at least ten by the addition of a few parts per million of beryllium of zinc ions. A study by Hudson and Bacon (1958) compares the effectiveness of these and other additives. The reasons for the inhibiting effects are not known.

REFERENCES

ABELÉS, F. (ed) (1972) "Optical Properties of Solids", North Holland, Amsterdam, 1062 pp.

ABOU-EL-LEIL, M., HEASLEY, J. and OMAR, M.H. (1978) Physics Chem. Glasses 19, 37-40.

ALLAN, W.B. (1973) "Fibre Optics. Theory and Practice", Plenum Press, New York, 247 pp.

ANDREWS, A.I. (1935) "Porcelain Enamels", Garrard Press, Champaign, Ill., 633 pp.

BACON, F.R. (1968) Glass Ind. 49, 438-46, 494-9, 544-9.

BAMFORD, C.R. (1976) Physics Chem. Glasses 17, 209-13.

BAMFORD, C.R. (1977) "Colour Generation and Control in Glass", Elsevier, Amsterdam, 224 pp.

BANERJEE, S. and PAUL, A. (1974) J. Am. Ceram. Soc. 57, 286-90.

BARNOSKI, M.K. (ed) (1976) "Fundamentals of Optical Fiber Communications", Academic Press, New York, 227 pp.

BARTON, A.F.M. (1971) Rev. pure appl. Chem. 21, 49-66.

BATES, T. (1962) in "Modern Aspects of the Vitreous State", Vol. 2 (ed. J.D. Mackenzie), 195-224, Butterworth, London.

BAYNTON, P.L., RAWSON, H. and STANWORTH, J.E. (1956) Trans. electrochem. Soc. 104, 237-40.

BEATTIE, I.R. (1953) J. Soc. Glass Technol. 37, 240-8T.

BELL, R.J. (1972) Rep. Prog. Phys. 35, 1315-1409.

BLIZARD, J.R. and HOWITT, J.S. (1969/70) Glass Ind. 50, 573-5; 51, 16-18, 73-5.

BOOW, J. and TURNER, W.E.S. (1942) J. Soc. Glass Technol. 26, 215-37.

BOWLES, R. (1976) British Glass Industry Research Association Technical Report No. 52.

BRAY, P.J. and SILVER, A.H. (1960) in "Modern Aspects of the Vitreous State" Vol. I (ed. J.D. Mackenzie), 92-119, Butterworth, London.

BRAY, P.J. and O'KEEFE, J.G.O. (1963) Physics Chem. Glasses, 4, 37-46.

BREARLEY, W. and HOLLOWAY, D.G. (1963) Physics Chem. Glasses, 4, 69-75.

BREWSTER, G.J., KREIDL, N.J. and PETT, T.G. (1947) J. Soc. Glass Technol. 31, 153-69.

BUDD, S.M. and KIRWAN, J.J. (1962) in "Advances in Glass Technology" Vol. 1, 527-40, Plenum Press, New York.

BUDD, S.M. (1966) Physics Chem. Glasses $\underline{7}$, 210-3.

BUDD, S.M. and CORNELIUS, W.P. (1976) Glass Technol. $\underline{17}$, 54-9.

BURT, R.C. (1925) J. opt. Soc. Am. $\underline{11}$, 87-91.

CAMERON, B.M. and HORNE, M.H. (1964) Glass Technol. $\underline{5}$, 110-4.

CHARLES, R.J. (1958a) J. appl. Phys. $\underline{20}$, 1549-53.

CHARLES, R.J. (1958b) J. appl. Phys. $\underline{20}$, 1554-60.

CHARLES, R.J. (1958c) J. appl. Phys. $\underline{20}$, 1657-62.

CHARLES, R.J. (1965) J. Am. Ceram. Soc. $\underline{48}$, 432-3.

CHARLES, R.J. (1966) J. Am. Ceram. Soc. $\underline{49}$, 55-62.

CHARLES, R.J. (1973) Bull. Am. Ceram. Soc. $\underline{52}$, 673-80, 86.

CLARK, D.E., DILMORE, M.F., ETHRIDGE, E.C. and HENCH, L.L. (1976) J. Am. Ceram. Soc. $\underline{59}$, 62-5.

CORBET, A. (1976) Nuclear Eng. Int. $\underline{21}$, 63-6.

CORNELISSEN, J. and ZIJLSTRA, A.L. (1962) in "Symposium sur la Résistance Mécanique du Verre", 337-58, U.S.C.V., Charleroi.

COTTRELL, A.H. (1964) "The Mechanical Properties of Matter", Wiley, New York, 430 pp.

COUSEN, A. (1936) J. Soc. Glass Technol. $\underline{20}$, 418-27T.

COUSEN, A. (1937) J. Soc. Glass Technol. $\underline{21}$, 177-86T.

DAS, C.R. and DOUGLAS, R.W. (1967) Physics Chem. Glasses $\underline{8}$, 178-84.

DAVIES, H.A. (1976) Physics Chem. Glasses $\underline{17}$, 159-73.

DEEG, E. (1965) Proc. VIIth Int. Congr. Glass, Brussels, Paper 1.

DENTON, E.P., RAWSON, H. and STANWORTH, J.E. (1954) "Nature" $\underline{173}$, 1080-1.

DETTRE, R.H. and JOHNSON, R.E. (1969) J. Adhes. $\underline{1}$, 92-101.

DIETZEL, A. and BRÜCKNER, R. (1955) Glastech. Ber. $\underline{28}$, 455-67.

DIETZEL, A. and WICKERT, H. (1956) Glastech. Ber. $\underline{29}$, 1-4.

DITCHBURN, R.W. (1963) "Light" 2nd End., Blackie, London, 833 pp.

DITCHBURN, R.W. (1976) "Light" 3rd Edn., Blackie, London, 775 pp.

DOREMUS, R.H. (1964) J. chem. Phys. 40, 2389-96.

DOREMUS, R.H. (1966) J. chem. Phys. 42, 414-7.

DOREMUS, R.H. (1973) "Glass Science" Wiley-Interscience, New York, 349 pp.

DOUGLAS, R.W., ARMSTRONG, W.L., EDWARDS, J.P. and HALL, D. (1965) Glass Technol. 6, 52-5.

DOUGLAS, R.W. and FRANK, S. (1972) "A History of Glass Making". G.J. Foulis, Henley-on-Thames, 213 pp.

DOUGLAS, R.W. (1974) Proc. Xth Int. Congr. Glass, Tokyo, 1(1), 45-70.

DUFFY, J.A., INGRAM, M.D. and SOMMERVILLE, I.D. (1978) J. chem. Soc. Faraday Trans. I. 4, 1410-9.

EL-SHAMY, T.M.M. and DOUGLAS, R.W. (1967) J. Am. Ceram. Soc. 50, 1-7.

EL-SHAMY, T.M.M., LEWINS, J. and DOUGLAS, R.W. (1972) Glass Technol. 13, 81-7.

ENSS, J. (1928) Glastech. Ber. 5, 449-76.

ERNSBERGER, F.M. (1959) J. Am. Ceram. Soc. 42, 373-5.

ERNSBERGER, F.M. (1960) Proc. R. Soc. A257, 213-23.

ERNSBERGER, F.M. (1962) in "Advances in Glass Technology", 511-524, Plenum Press, New York.

ERNSBERGER, F.M. (1965) Mach. Des. 37, 191-6.

ESCHARD, G. and MANLEY, B.W. (1971) Acta electron. 14, 19-39.

ESPE, W. (1968) "Materials of High Vacuum Technology", Vol. 2, Pergamon, Oxford, 660 pp.

EUROPEAN PHARMACOPOEIA (1971) "Procedure for the Determination of the Hydrolytic Resistance of the Internal Surface of Glass Containers".

EVSTROPIEV, K.S. and IVANOV, A.O. (1963) in "Advances in Glass Technology", Vol. 2, 79-85, Plenum Press, New York.

FAJANS, K. and JOOS, G. (1924) Z. Phys., 23, 1-46.

FAJANS, K. and KREIDL, N.J. (1948) J. Am. Ceram. Soc. 31, 105-14.

FARADAY, M. (1830) Phil. Trans. R. Soc. 1-57.

FINE, M.E. (1964) "Phase Transformations in Condensed Systems", Macmillan, London, 133 pp.

295

FONTANA, E.H. (1970) Bull. Am. Ceram. Soc. 49, 594-7.

FRANK, F.C. and LAWN, B.R. (1967) Proc. R. Soc. A299, 291-306.

FROCHT, M.M. (1948) "Photo-elasticity" 2 Vols. John Wiley, New York.

GAISER, R.A., LYON, K.C. and SCHOLES, A.B. (1965) Ceramic Ind. 84, 96-100, 136-40.

GARDON, R. (1961) J. Am. Ceram. Soc. 44, 305-12.

GARFINKEL, H.M. and KING, C.B. (1969) Glass Ind. 50, 28-31, 74-6.

GARFINKEL, H.M. and KING, C.B. (1970) J. Am. Ceram. Soc. 53, 686-91.

GILMAN, J.J. (1975) Phys. Today 28, 46-53.

GLATHART, J.L. and PRESTON, F.W. (1968) Glass Technol. 9, 89-100.

GLIEMEROTH, G., KRAUSE, D. and NEUROTH, N. (1976) Schott Information, 2, 1-17.

GOLDSMITH, W. and TAYLOR, R.L. (1976) Exp. Mech. 16, 81-7.

GOSSINK, R.G. (1977) J. Non-Cryst. Solids 26, 113-57.

GRAF, J. and POLAERT, R. (1973) Acta electron. 16, 11-22.

GRAUER, O.H. and HAMILTON, E.H. (1950) J. Res. natn. Bur. Stand. 44, 495-502.

GRIFFITH, A.A. (1920) Phil. Trans. R. Soc. A221, 163-198.

GRISCOM, D.L. (1977) J. Non-Cryst. Solids 24, 155-234.

GUYER, E.M. (1969) Glass Ind. 50, 186-90, 214.

HAFNER, H.C., KREIDL, N.J. and WEIDEL, R.A. (1958) J. Am. Ceram. Soc. 41, 313-23.

HAGY, H.E. (1963) J. Am. Ceram. Soc. 46, 93-7.

HAGY, H.E. (1968) J. Can. Ceram. Soc. 37, LXV-LXVIII.

HAMILTON, B. and RAWSON, H. (1970) J. Mech. Phys. Solids 18, 127-47.

HAMILTON, B. and RAWSON, H. (1972) in "Amorphous Materials" (eds B. Ellis and R.W. Douglas) 523-30, Wiley-Interscience, London.

HAMILTON, E.H., GRAUER, O.H., ZABOWSKI, Z. and HAHNER, C.H. (1948) J. Am. Ceram. Soc. 31, 132-44.

HANNA, R. and SU, G.J. (1964) J. Am. Ceram. Soc. 47, 597-601.

HARPER, D.W. and BOULTON, G.B. (1969) in "Optical Instruments and Techniques" (ed J. Home Dickson), 177-88, Oriel Press.

HEADY, R.B. and CAHN, J.W. (1973) J. chem. Phys. 58, 896-910.

HENCH, L.L. (1977) J. Non-Cryst. Solids 25, 343-69.

HENCH, L.L. and SCHAAKE, H.F. (1972) in "Introduction to Glass Science" (eds L.D. Pye, H.J. Stevens and W.C. La Course), 583-659, Plenum Press, New York.

HENDERSON, S.T. and MARSDEN, A.H. (eds) (1972) "Lamps and Lighting", Arnold, London, 602 pp.

HENDRICKSON, J.R. and BRAY, P.J. (1972) Physics Chem. Glasses, 13, 43-49, 107-115.

HENSLER, J.R. and LELL, E. (1969) Proc. Annual Meeting I.C.G., Toronto, 51-7.

HERRING, A.P., DEAN, R.W. and DROBNICK, J.L. (1970) Glass Ind. 51, 316-22, 350-6, 394-9.

HILLIG, W.B. (1961) J. appl. Phys., 32, 741.

HILTON, A.R. (1966) Appl. Opt. 5, 1877-82.

HOLLAND, L. (1963) Glass Ind., 44, 191-5, 268-72, 296-7, 386-94, 410-2, 445-50, 465-6.

HOLLOWAY, D.G. (1973) "The Physical Properties of Glass", Wykeham Publications, London, 217 pp.

HOLMES, J.G. (1945) J. scient. Instrum. 22, 219-21.

HOOGENDORN, H. and SUNNERS, B. (1969) Bull. Am. Ceram. Soc. 48, 1125-7.

HOPPER, E.S. (1968) Am. Soft Drinks J. 30-

HOWELL, B.F., SIMMONS, J.H. and HALLER, W. (1975) Bull. Am. Ceram. Soc. 54, 707-9.

HUDSON, G.A. and BACON, F.R. (1958) Bull. Am. Ceram. Soc. 37, 185-8.

HUGHES, J.V. (1941) J. scient. Instrum. 18, 234-7.

HUGHES, K. and ISARD, J.O. (1972) in "Physics of Electrolytes", Vol. I, (ed J.H. Hladik) 351-400, Academic Press, London.

INGLIS, C.E. (1913) Trans. Instn. nav. Archit. 55, 219-41.

INTERNATIONAL COMMISSION ON GLASS (1965, 1972, 1973) "Chemical Durability of Glass. A Review of the Literature", I.C.G., Brussels.

INTERNATIONAL COMMISSION ON GLASS (1970) "Viscosity-temperature Relations in Glass", I.C.G., Brussels

INTERNATIONAL COMMISSION ON GLASS (1977) "Electrical Properties of Glasses, Glass-Ceramics and Amorphous Solids. A Bibliography", I.C.G., Brussels.

INTERNATIONAL STANDARDS ORGANIZATION (1968a) "Determination of the Hydrolytic Resistance of Glass Grains at 98°C", I.S.O., 719.

INTERNATIONAL STANDARDS ORGANIZATION (1968b) "Determination of the Hydrolytic Resistance of Glass Grains at 121°C", I.S.O., 720.

INTERNATIONAL STANDARDS ORGANIZATION (1970) "Determination of the Resistance of Glass to Attack by 6NHCl at 100°C", I.S.O., 1776.

INTERNATIONAL STANDARDS ORGANIZATION (1977) "Glass Transfusion Bottles for Medical Use - Chemical Resistance", I.S.O., 3825.

ISARD, J.O. and DOUGLAS, R.W. (1949) J. Soc. Glass Technol. 33, 289-335T.

ISARD, J.O. (1959) J. Soc. Glass Technol. 43, 113-23T.

ISARD, J.O. (1969) J. Non-Cryst. Solids 1, 235-61.

IVANOV, A.O. (1963) in "The Structure of Glass" Vol. 7, (ed E.A. Porai-Koshits) 100-2, Consultants Bureau, New York.

IVANOV, A.O., EVSTROPIEV, K.S. and DOROKHOVA, M. (1965) in "The Electrical Properties of Glass" (ed O.V. Mazurin) 86-7, Consultants Bureau, New York.

IZUMITANI, T. and NAKAGAWA, K. (1965) Proc. Vth Int. Congr. Glass, Brussels, Paper 5.

IZUMITANI, T. and MATSUURA, T. (1967) in "Symposium on Coloured Glasses", Prague, I.C.G., Brussels, 98-113.

JAKOB, M. (1949) "Heat Transfer" 2 vols, Wiley, New York.

JAMES, P.F. (1975) J. Mater. Sci. 10, 1802-25.

JOHNSON, K.D.B. and MARPLES, J.A.C. (1979) "Glasses and Ceramics for Immobilisation of Radioactive Wastes for Disposal", H.M.S.O., London.

JOHNSON, K.L., O'CONNOR, J.J. and WOODWARD, A.C. (1973) Proc. R. Soc. A334, 95-117.

JOHNSTON, W.D. (1964) J. Am. Ceram. Soc. 47, 198-201.

JOHNSTON, W.D. (1965) J. Am. Ceram. Soc. 48, 184-90.

JONES, G.O. (1971) "Glass" 2nd edn. Chapman and Hall, London.

298

KAO, K.C. and HOCKHAM, G.A. (1966) Proc. I.E.E. 113, 1151-8.

KAPANY, N.S. (1960) Scient. Am. 203, 72-81.

KAPANY, N.S. (1967) "Fiber Optics. Principles and Applications", Academic Press, New York, 429 pp.

KATANYAN, K.A., SAAKYAN, K.S. and AVETISYAN, E.M. (1965) in "Electrical Properties and Structure of Glass" (ed O.V. Mazurin) 81-3, Consultants Bureau, New York.

KAY, S.E. (1973) Chemy. Ind. (Dec.), 1086-94.

KELLY, A. (1973) "Strong Solids" 2nd Edn. Clarendon Press, Oxford, 285 pp.

KERPER, M.J. and SCUDERI, T.G. (1964) Proc. Am. Soc. Test. Mater. 64, 1037-43.

KERPER, M.J. and SCUDERI, T.G. (1964a) Bull. Amer. Ceram. Soc. 43, 622-5.

KIRBY, P.L. (1956) J. Soc. Glass Technol. 40, 445-61T.

KOHL, W.H. (1972) "Handbook of Materials and Techniques for Vacuum Tubes", Reinhold, New York, 623 pp.

KORDES, E. (1956) Z. phys. Chem. 8, 318-41.

KORDES, E. (1965) Glastech. Ber. 38, 242-9.

KRAUSE, J.T. and KURKJIAN, C.R. (1968) J. Am. Ceram. Soc. 51, 226-7.

KROGH-MOE, J. (1965) Physics Chem. Glasses 6, 46-54.

KURKJIAN, C.R. (1963) Physics Chem. Glasses 4, 128-136.

LAHIRI, D., MUKHERJEE, B. and MAJUMDAR, R.N. (1974) Glastech. Ber. 47, 4-9.

LAKATOS, T., JOHNASSON, L-G., and SIMMINSKÖLD, B. (1972) Glass Technol. 13, 88-95.

LAKATOS, T., JOHANSSON, L-G., and SIMMINSKÖLD, B. (1973) Glastek. Tidsk. 28, 69-73.

LARNER, L.J., SPEAKMAN, K. and MAJUMDAR, A.J., J. Non-Cryst. Solids 20, 43-74.

LENGYEL, B.A. (1966) "An Introduction to Laser Physics", Wiley, New York, 311 pp.

LEVIN, E.M., ROBBINS, C.R. and McMURDIE, C.F. (1964: Supp 1; 1969: Supp 2, 1975) "Phase Diagrams for Ceramists", American Ceramic Society, Columbus, Ohio.

LILLIE, H.R. (1931) J. Am. Ceram. Soc. 14, 502-11.

LILLIE, H.R. (1933) J. Am. Ceram. Soc. 16, 619-631.

LILLIE, H.R. (1954) J. Am. Ceram. Soc. 37, 111-7.

LILLIE, H.R. and RITLAND, H.N. (1954) J. Am. Ceram. Soc. 37, 466-73.

LISTER, R.D. (1961) Glass Technol. 2, 186-91.

LORENTZ, H.A. (1906) "Theory of Electrons", Teubner, Leipzig.

MACEDO, P.B. and LITOVITZ, T.A. (1965) J. chem. Phys. 42, 245-56.

MACKENZIE, J.D. (1974) Proc. Xth Int. Congr. Glass, Kyoto, I.4, 71-81.

MASSON, C.R. (1977) J. Non-Cryst. Solids 25, 1-42.

MATSUMOTO, T. and MADDEN, R. (1975) Mat. Sci. and Engng. 19, 1-24.

MAURER, R.D. (1958) J. appl. Phys. 29, 1-8.

MAURER, R.D. (1976) Proc. I.E.E. 123, 581-5.

MAURER, R.D. (1977) J. Non-Cryst. Solids 25, 323-42.

MAZURIN, O.V. and BRAILOVSKAYA, R.V. (1960) Soviet Phys. Solid St. 2, 1341-5.

MAZURIN, O.V., STREL'TSINA, M.V. and SHVAIKO-SHVAIKOVSKYA, T.P. (1975) "The Properties of Glasses and Glass-forming Melts" 3 Vols. Izdatel'stvo 'Nauka' Leningradskoe Odelenie, Leningrad.

McELROY, J.L. (1975) Quarterly Progress Report April-June 1975. Battelle Pacific North West Laboratories. BNWL-1932.

McMILLAN, P.W. (1964) "Glass Ceramics" Academic Press, London, 229 pp. (2nd Edn. 1979).

McMILLAN, P.W. (1976) Physics Chem. Glasses, 17, 193-204.

McSWAIN, B.D., BORRELLI, N.F. and SU G-J. (1963) Physics Chem. Glasses 4, 1-10.

MEINECKE, G. (1959) Glas-Email-Keramo-Tech. 10, 209-12.

MEISTRING, R., FRISCHAT, G.H. and HENNICKE, H.W. (1976) Glastech. Ber. 49, 60-6.

METCALFE, A.G. and SCHMITZ, G.K. (1964) Proc. Am. Soc. Test Mater. 64, 1075-93.

MILLER, S.E., MARCATALI, E.A.J. and LI, T. (1973) Proc. I.E.E. 120, 1703-51.

MILNE, A.J. (1952) J. Soc. Glass Technol. 36, 275-86T.

MOCHEL, E.L., NORDBERG, M.N. and ELMER, T.E. (1966) J. Am. Ceram. Soc. 49, 585-9.

MONACK, A.J. and BEETON, E.E. (1939) Glass Ind. 20, 127-32, 185-91, 223-8, 257-62.

MOODY, B.E. (1963) "Packaging in Glass" 1st Edn., Hutchinson, London, 304 pp.

MOODY, B.E. (1977) "Packaging in Glass" 2nd Edn., Hutchinson-Benham, London, 383 pp.

MOREY, G.W. (1937) U.S. Patent 2,150,694.

MOREY, G.W. (1954) "The Properties of Glass" 2nd Edn. Reinhold, New York, 591 pp.

MORIAN, H. (1973) Schott Information (3) 16-18.

MOTT, N.F. (1968) J. Non-Cryst. Solids 1, 1-17.

MOTT, N.F. (1969) Phil. Mag. 19, 835-52.

MOTT, N.F. (1977) Contemp. Phys. 18, 225-45.

MOTT, N.F. (1978) J. Non-Cryst. Solids 28, 147-58.

MOTT, N.F. and DAVIS, E.A. (1971) "Electronic Processes in Non-crystalline Materials", Oxford (2nd Edn. 1979).

MOULD, R.E. (1952) J. Am. Ceram. Soc. 35, 230-5.

MOULD, R.E. and SOUTHWICK, R.D. (1959) J. Am. Ceram. Soc. 42, 542-7, 582-92.

MURGATROYD, J.B. (1942) J. Soc. Glass Technol. 26, 155-71T.

NAPOLITANO, A. and HAWKINS, E.G. (1964) J. Res. natn. Bur. Stand. 68A, 439-48.

NAPOLITANO, A., MACEDO, P.B. and HAWKINS, E.G. (1965) J. Res. natn. Bur. Stand. 69A, 449-55.

NATH, P. and DOUGLAS, R.W. (1965) Physics Chem. Glasses 6, 197-202.

NEWTON, R.G. (1982) "The Deterioration and Conservation of Painted Glass: A Critical Bibliography". Corpus Vitrearum Medii Aevi Great Britain - Occasional Papers II. Oxford U.P. 1982

OBLAD, A.G. and NEWTON, R.F. (1937) J. Am. chem. Soc. 59, 2495-9.

OLDFIELD, L.F. (1964) Glass Technol. 5, 41-50.

OLSHANSKY, R. and MAURER, R.D. (1976) J. appl. Phys., 47, 4497-9.

ORGEL, L.E. (1960) "An Introduction to Transition Metal Chemistry" Methuen, London, 180 pp.

OROWAN, E. (1949) Rep. Prog. Phys. 12, 185-232.

OROWAN, E. (1955) Weld. J. 34, 157-6Os.

ORR, L. (1972) Materials Res. Standards, 12, 21-3, 47.

OVSHINSKY, S.R. (1968) Phys. Rev. Lett. 21, 1450-3.

OWEN, A.E. (1963) in "Progress in Ceramic Science" Vol. 3 (ed J.E. Burke) 78-198, Pergamon, London.

OWEN, A.E. (1970) Contemp. Phys. 11, 227-86.

OWEN, A.E. (1977) J. Non-Cryst. Solids 25, 371-423.

OWEN, A.E. and DOUGLAS, R.W. (1959) J. Soc. Glass Technol. 43, 159-78T.

OWEN, A.E. and SPEAR, W.E. (1976) Physics Chem. Glasses 17, 174-92.

PANTANO, C.G., DOVE, D.B. and OKADA, G.V. (1975) J. Non-Cryst. Solids 19, 41-53.

PANTELIDES, S.T. and HARRISON, W.A. (1976) Phys. Rev. B13, 2667-91.

PARKE, S. (1974) in "The Infra-red Spectra of Minerals" (ed V.C. Farmer) 483-514, The Mineralogical Society, London.

PARMALEE, C.S. (1951) "Ceramic Glazes" 2nd Edn. Industrial Publications, Chicago, 322 pp.

PARTRIDGE, J.H. (1949) "Glass-to-Metal Seals", Society of Glass Technology, Sheffield, 238 pp.

PATEK, K. (1970) "Glass Lasers" Iliffe, London, 217 pp.

PAUL, A. (1970) Physics Chem. Glasses 11, 46-52.

PAUL, A. (1971) Trans. Indian Ceram. Soc. 30, 73-8.

PAUL, A. (1974) J. Non-Cryst. Solids 15, 517-25.

PAUL, A. (1975) J. Mater. Sci. 10, 415-21.

PAUL, A. (1977) J. Mater. Sci. 12, 2246-68.

PAUL, A. and DOUGLAS, R.W. (1965) Physics Chem. Glasses 6, 212-5.

PAUL, A. and DOUGLAS, R.W. (1966) Physics Chem. Glasses 7, 1-13.

PAUL, A. and DOUGLAS, R.W. (1968) Physics Chem. Glasses 9, 21-6.

PEARSON, A.D., DEWALD, J.F., NORTHOVER and PECK, W.F. (1962) in "Advances in Glass Technology" Vol. I, 357-65, Plenum Press, New York.

PEDDLE, C.J. (1937) J. Soc. Glass Technol. 21, 187-95T.

PERSSON, H.R. (1962) Glass Technol. 3, 17-35.

PHILLIPP, H.R. (1966) Solid St. Commun. 4, 73-5.

PHYSICAL PROPERTIES COMMITTEE (1956) J. Soc. Glass Technol. 40, 83-104P.

PLUMMER, W.A. and HAGY, H.E. (1968) Appl. Opt. 7, 825-31.

POOLE, J.B. (1967) Glass Ind. 48, 129-36.

PRESTON, F.W. (1931) J. Am. Ceram. Soc. 14, 419-27.

PRESTON, F.W. (1939) Bull. Am. Ceram. Soc. 18, 35-60.

PROCTOR, B. (1962) Physics Chem. Glasses 3, 7-27.

PYE, L.D., STEVENS, H.J. and LACOURSE, W.C. (1972) "Introduction to Glass Science", Plenum Press, New York, 722 pp.

RAM, A. (1961) Proc. Nat. Inst. Sci. India, 27A, 531-67.

RAM, A. and PRASAD, S.N. (1962) in "Advances in Glass Technology", Vol. I, 256-9, Plenum Press, New York.

RANA, M.A. and DOUGLAS, R.W. (1961) Physics Chem. Glasses 2, 179-95.

RAWSON, H. (1965) Physics Chem. Glasses 6, 31-4.

RAWSON, H. (1967) "Inorganic Glass-forming Systems" Academic Press, London, 317 pp.

RAWSON, H. (1974) Phys. Technol. 5, 91-114.

RAWSON, H. (1977) J. Non-Cryst. Solids 26, 1-25.

RITTER, J.E. and SHERBOURNE, C.L. (1971) J. Am. Ceram. Soc. 54, 601-5.

ROCKETT, J.J., FOSTER, W.R. and FERGUSON, R.G. (1965) J. Am. Ceram. Soc. 48, 329-31.

SAVAGE, J.A. and NIELSON, S. (1965) Infrared Phys. 5, 195-204.

SAVAGE, J.A., WEBBER, P.J. and PITT, A.N. (1977) Appl. Opt. 16, 2938-41.

SCHARDIN, H. (1959) in "Fracture" (eds B.L. Averbach, D.K. Felbeck, G.T. Hahn and D.A. Thomas) Wiley, New York.

SCHOLES, S.R. (1975) "Modern Glass Practice", 7th Edn.,
Cahners Books, Boston, Mass., 493 pp.

SCHOLZE, H. (1959) Glastech. Ber. 32, 81-8, 142-52, 278-81,
314-20, 381-6.

SCHOLZE, H. (1965) "Glas. Natur, Struktur und Eigenschaften",
Vieweg, Brunswick, 370 pp.

SCHOLZE, H. (1977) "Glas. Natur, Struktur und Eigenschaften",
Springer, Berlin, 342 pp.

SCOTT, B. and RAWSON, H. (1973) Glass Technol. 14, 115-25.

SEITZ, F. (1940) "The Modern Theory of Solids", McGraw-Hill,
New York, 698 pp.

SHAND, E.B. (1958) "Glass Engineering Handbook", 2nd Edn.,
McGraw Hill, New York, 484 pp.

SHAND, E.B. (1959) J. Am. Ceram. Soc., 42, 474-7.

SHARTSIS, L. and SPINNER, S. (1951) J. Res. natn. Bur. Stand.,
46, 176-94.

SHARTSIS, L., SPINNER, S. and CAPPS, W. (1952) J. Am. Ceram.
Soc., 35, 155-60.

SIEGER, J.S. (1975) J. Non-Cryst. Solids, 19, 213-20.

SIGEL, G.H. (1973) J. Non-Cryst. Solids, 13, 372-98.

SIMMONS, J.H., MILLS, S.A. and NAPOLITANO, A. (1974) J. Am.
Ceram. Soc., 57, 109-17.

SIMMONS, J.H. (1977) J. Non-Cryst. Solids, 24, 77-88.

SIMPSON, H.E. (1951) Bull. Am. Ceram. Soc., 30, 41-5.

SIMPSON, H.E. (1959) J. Am. Ceram. Soc., 42, 337-43.

SMITH, G.P. (1967) J. mat. Sci., 2, 139-52.

SMITH, R.A. (1961) "Wave-Mechanics of Crystalline Solids",
Chapman and Hall, London, 473 pp.

SNITZER, E. (1961) Phys. Rev. Lett., 7, 444-6.

SNITZER, E. (1964) in "Quantum Electronics" (eds P. Grivet and
N. Bloembergen), 999-1019, Columbia University Press.

SNITZER, E. (1966) Appl. Opt. 5, 1487-99.

SNITZER, E. (1973) Bull. Am. Ceram. Soc. 52, 516-25.

STANEK, J. (1977) "Electric Melting of Glass", Elsevier,
Amsterdam, 391 pp.

STANWORTH, J.E. (1952) Nature, 169, 581.

STEVELS, J.M. (1948) Verres Réfract. 2, 1-14.

STEVELS, J.M. (1953) Proc. 11th Int. Congr. pure appl. Chem., 5, 519-22.

STEVELS, J.M. (1957) Handbuch der Physik, 20, 350-91.

STOOKEY, S.D. and MAURER, R.D. (1961) in "Progress in Ceramic Science, Vol. 2, (ed. J.E. Burke) 77-101, Pergamon, London.

STROUD, J.S. and LELL, E. (1971) J. Am. Ceram. Soc. 54, 554-5.

SWIFT, H.R. (1947) J. Am. Ceram. Soc. 30, 170-4.

SYKES, R.F.R. (1965) Glass Technol. 6, 178-83.

TAKAYAMA, S. (1976) J. Mater. Sci. 11, 164-85.

TAYLOR, H.E. (1957) J. Soc. Glass Technol. 41, 350-82T.

TAYLOR, H.E. (1959) J. Soc. Glass Technol. 43, 124-46T.

TEAGUE, J.M. and BLAU, H.H. (1965) J. Am. Ceram. Soc. 39, 229-52.

TENQUIST, D.W., WHITTLE, R.M. and YARWOOD, J. (1969, 170) "University Optics" 2 Vols, Iliffe, London.

THOMAS, W.F. (1960) Physics Chem. Glasses 1, 4-18.

TICHANE, R.M. (1966) Glass Technol. 7, 26-9.

TICHANE, R.M. and CARRIER, G.B. (1961) J. Am. Ceram. Soc. 44, 606-10.

TICKLE, R.E. (1967) Physics Chem. Glasses 8, 101-24.

TIEDE, R.L. (1959) J. Am. Ceram. Soc. 62, 537-41.

TIMOSHENKO, S. and GOODIER, R.N. (1951) "Theory of Elasticity", McGraw-Hill, New York, 506 pp.

TIMOSHENKO, S. and WAINOWSKI-KRIEGER, S. (1959) "Theory of Plates and Shells", 2nd Edn., McGraw-Hill, New York, 580 pp.

TOOLEY, F.V. (1971) "Handbook of Glass Manufacture" 2 Vols, 2nd Edn., Books for Industry Inc., New York.

TRAP, H.J.L. (1971) Acta electron. 14, 41-77.

TRIER, W. (1961) J. Am. Ceram. Soc. 44, 339-45.

TSAI, C.R. and STEWART, R.A. (1976) J. Am. Ceram. Soc. 59, 445-8.

TSUCHIHASHI, S., KONISHI, A. and KAMAMOTO, Y. (1975) J. Non-Cryst. Solids, 19, 221-9.

TURNBULL, D. (1969) Contemp. Phys. 10, 473-88.

TURNBULL, D. and COHEN, M.H. (1958) J. chem. Phys. 29, 1049-54.

TURNER, W.E.S. and WINKS, F. (1928) J. Soc. Glass Technol. 12, 57-74T.

TURNER, W.E.S. and WINKS, F. (1930a) J. Soc. Glass Technol. 14, 84-109T.

TURNER, W.E.S. and WINKS, F. (1930b) J. Soc. Glass Technol. 14, 110-26T.

TURTON, G. and RAWSON, H. (1973) Glastech. Ber. 46, 28-33.

U.S. FLAT GLASS MANUFACTURING ASSOCIATION (1974) "Glazing Manual", Topeka, Kansas.

VOGEL, W. and GERTH, K. (1958) Glastech. Ber. 31, 15-28.

VOLF, M.B. (1961) "Technical Glasses", Pitman, London, 465 pp.

VON HIPPEL, A.R. (ed)(1954) "Dielectric Materials and Applications", Wiley, New York, 438 pp.

WAGNER, C. (1975) Metall. Trans. 6B, 405-9.

WALTERS, H.V. and ADAMS, P.B. (1975) J. Non-Cryst. Solids 19, 183-99.

WARNER, D. and RAWSON, H. (1978) J. Non-Cryst. Solids 29, 231-7.

WARREN, B.E. and MOZZI, R.L. (1969) J. appl. Crystallogr. 2, 164-72.

WASHINGTON, D., DUCHENOIS, V., POLAERT, R. and BEASLEY, R.M. (1971) Acta electron. 14, 201-24.

WEISER, K. (1976) Prog. Solid State Chem. 11, 403-445.

WELCH, M.A. and HUDSON, R.R. (1976) Glass Technol. 17, 257-62.

WERNER, A.J. (1968) Appl. Opt. 7, 837-43.

WESTMAN, A.E.R. (1960) in "Modern Aspects of the Vitreous State" Vol. I, (ed J.D. Mackenzie) 63-71, Butterworth, London.

WEYL, W.A. (1951) "Coloured Glasses", Society of Glass Technology, Sheffield, 541 pp.

WIEDERHORN, S.M. (1966) in "Materials Science Research" Vol. 3, (eds W.W. Kriegel and H. Palmour) 503-25, Plenum Press, New York.

WIEDERHORN, S.M. and BOLZ, L.H. (1970) J. Am. Ceram. Soc. 53, 543-8.

WIEDERHORN, S.M., JOHNSON, H., DINESS, A.M. and HEUER, A.H. (1974a) J. Am. Ceram. Soc. 57, 336-41.

WIEDERHORN, S.M., EVANS, A.G. and FULLER, E.R., (1974b) J. Am. Ceram. Soc. 57, 319-23.

WIEDERHORN, S.M. and EVANS, A.G. (1974) Int. J. Fracture, 10, 379-92.

WONG, J. and ANGELL, C.A. (1976) "Glass Structure by Spectroscopy", Dekker, New York, 864 pp.

WORK, R.N. (1951) J. Res. natn. Bur. Stand. 47, 80-6.

WRIGHT, W.D. (1964) "The Measurement of Colour", 3rd Edn., Hilger and Watts, London, 291 pp.

YATES, J.A. (1974) Glass Technol. 15, 21-7.

YOUNG, C.G. (1969) Proc. I.E.E., 57, 1267-89.

ZACHARIASEN, W.H. (1932) J. Am. chem. Soc. 54, 3841-51.

ZIJLSTRA, A.L. (1962) in "Symposium sur la Résistance Mécanique du Verre", 135-204, U.S.C.V., Charleroi.

ZIJLSTRA, A.L. and BURGRAAF, A.J. (1968) J. Non-Cryst. Solids 1, 49-68, 163-85.

ZIJLSTRA, A.L. and de GROOT, J. (1962) in "Symposium sur la Résistance Mécanique du Verre" 359-76, U.S.C.V., Charleroi.

ZINNGL, H. and SIMMINSKÖLD, B. (1967) Glastek. Tidsk. 22, 59-64.

AUTHOR INDEX

SUBJECT INDEX